D0291896

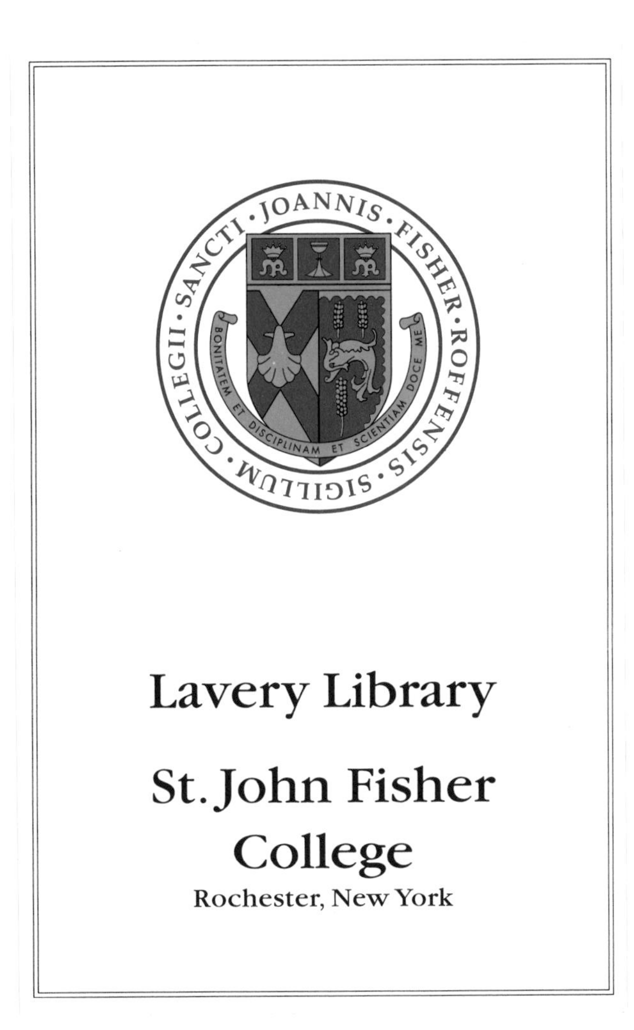
SIGILLUM · COLLEGII · SANCTI · JOANNIS · FISHER · ROFFENSIS ·
BONITATEM ET DISCIPLINAM ET SCIENTIAM DOCE ME

HIDING IN THE MIRROR

HIDING IN THE MIRROR

ALSO BY LAWRENCE M. KRAUSS

Atom: A Single Oxygen Atom's Journey from the Big Bang to Life on Earth . . . and Beyond

Quintessence: The Mystery of the Missing Mass

Beyond Star Trek: From Alien Invasions to the End of Time

The Physics of Star Trek

Fear of Physics: A Guide for the Perplexed

The Fifth Essence: The Search for Dark Matter in the Universe

HIDING IN THE MIRROR

HIDING IN THE MIRROR

THE MYSTERIOUS ALLURE OF EXTRA DIMENSIONS, FROM PLATO TO STRING THEORY AND BEYOND

Lawrence M. Krauss

VIKING

VIKING
Published by the Penguin Group
Penguin Group (USA) Inc., 375 Hudson Street, New York, New York 10014, U.S.A.
Penguin Group (Canada), 90 Eglinton Avenue East, Suite 700,
Toronto, Ontario, Canada M4P 2Y3 (a division of Pearson Penguin Canada Inc.)
Penguin Books Ltd, 80 Strand, London WC2R 0RL, England
Penguin Ireland, 25 St. Stephen's Green, Dublin 2, Ireland (a division of Penguin Books Ltd)
Penguin Books Australia Ltd, 250 Camberwell Road, Camberwell, Victoria 3124, Australia
(a division of Pearson Australia Group Pty Ltd)
Penguin Books India Pvt Ltd, 11 Community Centre, Panchsheel Park,
New Delhi - 110 017, India
Penguin Group (NZ), Cnr Airborne and Rosedale Roads, Albany,
Auckland 1310, New Zealand (a division of Pearson New Zealand Ltd)
Penguin Books (South Africa) (Pty) Ltd, 24 Sturdee Avenue, Rosebank,
Johannesburg 2196, South Africa

Penguin Books Ltd, Registered Offices:
80 Strand, London WC2R 0RL, England

First published in 2005 by Viking Penguin,
a member of Penguin Group (USA) Inc.

10 9 8 7 6 5 4 3 2 1

Figure on page 67: Brendan Crill, The Boomerang Collaboration.

All other drawings by the author.

ISBN 0-670-03395-2
CIP data available

Printed in the United States of America
Set in Baskerville Book with Helvetica Neue
Designed by Daniel Lagin

For my mother . . . at last!

There is a dimension, beyond that which is known to man.
It is a dimension as vast as space
and as timeless as infinity.
It is the middle ground between light and shadow,
between science and superstition,
and it lies between the pit of man's fears
and the summit of his knowledge.
This is the dimension of imagination.

–Rod Serling, *The Twilight Zone*

CONTENTS

HIDING IN THE MIRROR

REMINISCENCE
A DIMENSIONAL LOVE AFFAIR

Two parents wake in the middle of the night to sounds of their daughter's crying out in the distance. The father rushes to her bedroom and finds her missing. He frantically searches everywhere, slowly coming to the grim realization that she is gone. His wife runs into the room soon afterward, overcome with panic. At his wit's end, he dashes out to the living room and picks up the phone and calls a neighbor. He returns to his wife and, in words that are probably unique in the history of television, tells her:

"Bill's coming over. He's a physicist! He ought to be able to help!"

Forty-two years ago, when I was very young, a *Twilight Zone* episode called "Little Lost Girl" scared the living daylights out of me. Touching on every child's fear of being separated from the safety of parents and home, the episode told the story of a little girl who falls into another dimension.

When I first thought about writing a book that might focus on our love affair with extra dimensions, "Little Lost Girl" came immediately to mind, although I confess I had no memory of the episode's title or when it aired. After a short bit of research on the Web, I was able to locate it, and a few days later, along with forty-two other episodes I had to buy in a *Twilight Zone* boxed set, it arrived at my door.

That night I placed the DVD into my computer and relived my childhood trauma. The eerie thing was that I remembered everything about the episode . . . *except for the physicist*! But suddenly, upon hearing that line of dialogue, a rush of memories came flooding back.

Of course! The physicist was the hero of the episode. He came over in the middle of the night, discovered and traced out the "portal" in the wall through which the small child and her dog had wandered, guided her father through the gap, and ulitimately reached through and saved the father and the terrified duo moments before this door to another dimension closed forever.

I now vividly remember (or I think I remember) being struck by how exotic and powerful Bill the physicist's knowledge seemed, and how much respect this knowledge engendered in his frightened neighbor. I, too, wanted one day to be privy to such secrets, and to explain them. I wanted to be the one whom people in distress knew they could count on. In short, the physicist-superhero!

Alas, I have been a physicist for over twenty years now, and except for some students every now and then the night before an exam, no one has sought out my physics expertise when in distress. Nevertheless, I sometimes wonder if I write books such as this to fulfill my desire to provide what Bill had offered his neighbors: insights that physics has revealed about universal human mysteries, such as from whence we came, and what may lie beyond the darkness of the night. Some people seek solace through the spirit, but for others it comes through knowledge.

As Rod Serling, the creator of the *Twilight Zone,* observed in his weekly introductions beginning in 1959, the human imagination can create whole universes into which we can travel via the depths of despair or the peaks of ecstasy. Ultimately our continuing intellectual fascination with extra dimensions may tell us more about our own human nature than it does about the universe itself.

We all yearn to discover new realities hidden just out of sight. So much so, that we have continually reinvented them throughout human history, whenever the world of our experience has seemed lacking. But this does not necessarily mean that all of these worlds beyond our direct experience are unreal. There are scientists today who truly expect to discover the existence of extra dimensions and perhaps even extra universes in our lifetime.

I originally began this book because I wanted to explore the unique cultural and scientific legacy that has led to our current fascination with exotic new realms that may lie hidden in the mirror. But I never guessed that the voyage could be so personal. I now realize that with seven words heard in a television program some forty-two years ago, my own future may have been determined.

CHAPTER 1
THE PRIVILEGE TO LIVE IN SPACE?

I call our world Flatland, not because we call it so, but to make its nature clearer to you, my happy readers, who are privileged to live in Space.

So begins perhaps the most famous mathematical romance ever written. Penned in 1884, twenty-one years before Albert Einstein revolutionized our notions of space and time, under the pseudonym "A. Square" by the clergyman and Shakespearean scholar Edwin A. Abbott, *Flatland* was a poignant tale told by a wistful two-dimensional being who had just discovered the miraculous existence of three-dimensional space and longed to enjoy it. The unhappy hero of this saga urged us lucky Spacelanders to recognize the beauty of the higher-dimensional universes that he thus envisaged.

At around the same time that Abbott was writing *Flatland,* a lonely and tragic artist on the Continent was imagining another universe beyond the realm of our perception. Vincent Van Gogh was a tortured genius who is said to have sold but a single painting in his lifetime. Yet you cannot walk the streets of Amsterdam today without seeing reproductions in storefront windows of his haunting self-portraits or his landscapes with yellow skies and blue earth. In 1882, he wrote to his brother, who was his sole supporter, "I know for certain that I have a feeling for color, and shall acquire more and more." Through his paintings Van Gogh freed our minds from the "tyranny" of color, daring us to imagine everyday objects in a completely different way, and thereby demonstrating that exotic realities could be discovered in even the otherwise most ordinary things. His paintings are haunting not because they are so bizarre but because they are just bizarre enough to capture the essence of reality while at the same time forcing us to reexamine what exactly reality is.

These are the luxuries of art and literature: to create imaginary worlds that cause us to reconsider our place within our own world. Science has comparable impact. It, too, unveils different sorts of hidden worlds, but ones that we hope might also actually exist and, most importantly, can be measured. Nevertheless, the net result is the same: In the end we gain new insights into our own standing in the universe.

All of these creative human activities reflect the essence of human imagination, the spark that raises our existence from the mundane to the extraordinary. If we couldn't imagine the world as it might be, it is possible that the world of our experience would become intolerable.

Such imagination almost defines what it means to be human. Fourteen thousand years ago, in what is now France, a remote Ice Age ancestor took a walk with a young child into what many of us today would think of as a dark and forbidding place. Deep in an underground cave the adult held the child's hand against a wall and blew pigment over it, leaving a shadowlike imprint of a tiny hand that remains to this very day.

We will never know the purpose of this adventure. Did it have some deep spiritual significance, or was it simply play? It certainly was not an everyday activity, as our Cro-Magnon ancestors did not tend to live in the deep recesses of caves such as this. Whatever its purpose, it represents something very special about humans that clearly differentiates us from our closest relatives on the evolutionary tree.

I am not speaking here about art per se. Rather, I am addressing the deeper, symbolic sense of self that art reflects. The notion that the imprint on a wall might permanently record the presence of two individuals in the cave that day implies not only a recognition of their own existence, but also their desire to preserve some aspect of it against the vicissitudes of a dangerous world. For with a sense of self comes a sense of everything that *isn't* self, of the "unknown possibilities of existence," as the godlike alien Q on *Star Trek* once described it.

That even earlier humans pondered such unknown possibilities is testified to by the existence of artistic renderings that predate the French cave art by at least eighteen thousand years. In a cave at a site called Hohlenstein-Stadel, in what is now Germany, a foot-tall figure of a standing human was discovered. No less striking than the skill of the artist who created it is the subject matter: This figure has the head of a lion, not a man.

Did this early carving represent some primal notion of a deity? Or did it merely represent the recognition that if lions existed, and humans existed, then somewhere, some exotic combination of the two might exist? Of course, here again we shall probably never know what motivated our ancestral carver, but whatever its purpose the figure reflects an artistic imagining of the possibilities inherent either in this world or in one beyond it.

In the three hundred centuries that have passed since this figure was created, human civilization, and human imagination, have evolved considerably. But there remains a fundamental connection between our modern efforts and these first, tentative steps: When we imagine the world beyond our experience, we are digging deep into our own psyches.

In the famous *Twilight Zone* quote with which I began this book, Rod Serling argued that imagination is the middle ground between science and superstition. With that in mind, the central question becomes: To what extent do our imaginings reflect our own predilections, and to what extent might they actually mirror reality?

If we can directly test our imaginings against the weight of observation and experiment, then the answer is easy. But what if we cannot? When certain notions persist, in many cultures and many times, are they merely hardwired in our brains? Or perhaps, even if they are, is it because we are the products of a natural world that incorporates them?

One such notion will be the focus of this book: the longstanding love affair of the human intellect with the idea that there is far more "out there" than meets the eye. Science has, of course, validated this notion. Whole new realms of the physical world have been exposed by the spectacular scientific developments of the nineteenth and twentieth centuries.

But in the present context I mean something more literally "out there." Could space itself extend beyond the bounds of our experience, and can there be whole new dimensions of space just out of reach of our senses? It is difficult to disagree with Serling that imagination adds an extra dimension to the human experience. Still, the question remains: Is a fifth–or even an eleventh, or twenty-sixth–dimension purely imaginary?

What if extra dimensions exist but they remain hidden from even the most sophisticated detectors? Can our imaginations alone enable us to pierce nature's veil to discover them?

This very question drove the most famous of all philosophers in Western history to write a tale about a two-dimensional world as an allegory for our own limited understanding of reality. Twenty-five hundred years ago, in his most famous set of Dialogues, *The Republic,* Plato invented the allegory of a cave to describe his belief in the possibility of uncovering hidden realities within all of the objects of our experience.

Plato envisaged our lives as being like those of individuals confined in shackles within a cave, unable to directly see the world of light beyond. These prisoners viewed all objects located outside the mouth of the cave via the shadows they cast on the cave's back wall. To the viewers, who had no other experience, the shadows themselves represented the real objects.

Imagine, says Plato, through his interlocutor, Socrates, what it would be like to be unchained and dragged out to the light outside. First, of course, the brilliant glare would be painful, and one would crave a return to the dark familiarity of the cave. Ultimately, however, the true wonder of the world would become intoxicating–so much so that a return to one's previous state of ignorant slavery would be unthinkable. And even if one did return, how would it be possible to communicate the truth without appearing mad to those who had no idea of it?

Plato argued, however, that this is precisely the responsibility of a true philosopher. He must be willing to forsake the comfort of his own safe vision of reality and embark on travels through frightening new terrains of the mind. But more important, he must not be content to remain in his ivory tower of learning, separate from the rest of the human race, but must be willing to return to the world of men, to attempt to educate those who govern the affairs of men in the true workings of the universe.

When Socrates was asked, in Plato's dialogue, how one could penetrate the fog that shields us from the true workings of reality, his response was particularly telling, especially in light of our current scientific perspective. The answer involved the study of abstractions–in particular, arithmetic, the science of numbers. Or, as he put it, "Number, then, appears to lead towards the truth."

The study of numbers, said Socrates, should be followed by, in successively lesser importance, the study of geometry, then astronomy–as far as it concerns the laws of motion–then perhaps harmony, the study of sound. Only through the study of abstractions of the mind–as he viewed these disciplines–could one release oneself from the chains that bind us all to the rigid world of our senses.

Plato's entreaties now appear hauntingly modern. If his own abstraction–via the two-dimensional shadows of three-dimensional objects–might open the minds of his contemporaries to the infinite possibilities of existence, what mysteries might modern mathematical excursions unveil? Perhaps this spirit supplemented Abbott's desire to create a piece of social satire when he penned *Flatland.*

Indeed, the twentieth-century British mathematician and philosopher Bertrand Russell, in his *Study of Mathematics,* echoed almost verbatim Plato's idealism about the hidden power of mathematics:

> Mathematics, rightly viewed, possesses not only truth, but supreme beauty . . . a beauty cold and austere, like that of sculpture, without appeal to any part of our weaker nature, without the gorgeous trappings of painting or music, yet sublimely pure, and capable of a stern perfection such as only the greatest art can show. The true spirit of delight, the exaltation, the sense of being more than Man, which is the touchstone of the highest excellence, is to be found in mathematics as surely as in poetry.

More recently we have become so accustomed to the superb predictive power of our mathematical descriptions of reality that it is easy to take this unexpected connection between human abstraction and the actual workings of the natural world for granted. Yet the mathematical physicist and Nobel laureate Eugene Wigner wrote a famous essay in 1960 entitled "The Unreasonable Effectiveness of Mathematics in the Natural Sciences." In it he mused about the remarkable success of mathematics as a description of natural phenomena, or, as he put it, "The enormous usefulness of mathematics in the natural sciences is something bordering on the mysterious and . . . there is no rational explanation for it."

It was precisely this latter fact–that the profound connection between mathematics and the natural world seems to be "a wonderful gift which we neither understand nor deserve," as Wigner put it–that led him to speculate further. Does the "uncanny usefulness of mathematical concepts" suggest that a perhaps wholly different mathematics from that we have exploited to describe nature might perform equally well? Namely, are our physical theories unique–do they represent some fundamental underlying reality about nature–or have we just chosen one of many different, possibly equally viable, mathematical frameworks within which to pose our questions? In this latter case, would the apparent underlying physical pictures corresponding to these other mathematical descriptions each be totally different?

Because we have made huge strides in our understanding of the nature of scientific theories in the intervening forty years since Wigner penned his essay, I believe we can safely say that the question he poses is no longer of any great concern to scientists. We understand precisely how different mathematical theories can lead to equivalent predictions of physical phenomena, because some aspects of the theory will be mathematically irrelevant at some physical scales and not at others. Moreover, we now tend to think in terms of "symmetries" of nature, what are reflected in the underlying mathematics. While this once again argues for the importance of mathematics in our understanding of nature, these symmetries themselves seem so fundamental that we expect that any theory that can produce correct predictions must reflect them. Thus, seemingly different mathematical formulations can really be understood to reflect identical underlying physical pictures.

There is also a flip side to the discussion regarding the unusual effectiveness of mathematics in describing nature. Not all novel mathematical notions that open new horizons for our imagination have correlatives in the natural world. If that were the case, science would be no more than searching for new mathematics.

The power of mathematics will play a large role in what follows, but when it comes to the relationship between our scientific imagination and reality, elegance or mathematical beauty is by itself not sufficient to generate fruitful science. What matters are results. That is why science isn't philosophy, and why nature holds the upper hand. As Richard Feynman once put it, science is "imagination in a straightjacket." In the end our theories rise and fall based on their successful ability to quantitatively predict the future. Imagination truly rises to the level of beauty in science when it allows one to make predictions about things that one may never have thought were predictable.

To return to Plato's cave, Socrates pointed out that the unfortunate soul who had literally seen the light would, when dragged back in the cave, appear at first to his former compatriots to be a lunatic. This does not, however, mean that all lunatics have seen the light. Every religious prophet in history, for example, from Moses to Jesus, from Mohammed to Joseph Smith has cloaked his or her revelations in language similar to Plato's. They all suggest that to see the true nature of the world, we merely have to remove the curtains in front of our eyes. But they cannot *all* be correct. There are different worlds behind each of their curtains.

Which brings us inevitably to another complementary aspect of the human experience that literally depends on the existence of another world: religion. It is perhaps not surprising that one of the most popular Christian writers of the the twentieth century, C. S. Lewis, produced a profoundly successful children's series, *The Chronicles of Narnia,* which literally exploited a whole new world hidden just under our noses in order to relay its highly allegorical epic saga. Lewis's Narnia was not like Tolkien's Middle Earth, located far, far away and long, long ago. Rather, it could be accessed simply by entering an old wardrobe located in a professor's cluttered house in the country. This was supposed to be some kind of magic, but it is in a fundamental sense not too different from Bill's portal through the fourth dimension that aired less than a decade later on the *Twilight Zone.*

Lewis's fantasy stems from a long tradition that indeed lies in that dimension that spans both science and superstition. There is undoubtedly a deep need within our psyches to believe in the existence of new realms where our hopes and dreams might be fulfilled, and our worst nightmares may lie buried.

Religion is the most obvious manifestation of this innate desire for a universe that may be far richer, and perhaps kinder and gentler, than our material existence belies. Nevertheless, while our longings for a deeper reality are in one sense deeply spiritual, they transcend the purely spiritual. They permeate all aspects of our culture, including the pursuit of science.

In order to separate science from superstition, we need to recognize that, like Fox Mulder in *The X-Files,* we all *want* to believe. Forcing our beliefs to conform to the realities of nature, however, rather than the other way around, is much more difficult and is really, in my opinion, one of the greatest gifts that science can provide our civilization.

The process by which this transformation from imagination to science is made is not always clear-cut, especially when we are embroiled in the middle of it as we certainly are now, at least as far as the possibility of new small or large extra dimensions in nature is concerned. This book will in part provide a timely snapshot of where we are now: of the physical and mathematical motivations for our speculations, the sudden rushes of clarity, and the many frustrating red herrings and dashed expectations. The picture that is emerging is far from being in focus, unlike much of what one might read in the popular press. But not knowing all of the answers, and perhaps more importantly, knowing that one does not know all of the answers, is what keeps the search exciting.

We shall encounter diverse manifestations, developed over several centuries—in art, literature, and science—of the idea that the three dimensions of space that we experience are not all there is. But this topic has in recent years taken on a special urgency, which is why I believe it is worth relating at this time, in an honest way, to a broader audience. Dramatic new theoretical ideas seem to suggest the existence of many extra dimensions, and scientists are at this very moment struggling to determine if they have any relation to the real world.

It is worth stressing this last point. Too often in the media, speculative ideas are treated on the same footing as well-tested ones. As a result, it is sometimes hard to tell the difference between them. This is particularly unfortunate when firmly grounded ideas that are known to accurately describe the physical world (such as evolution and the big bang) are passed off as mere theoretical whims of a group of partisan scientists. One of the most useful tasks a popular exposition of science at the forefront can achieve, it seems to me, is clearly differentiate that which we know yields an accurate description of nature on some scale from those things we have reason to *suspect* one day might do so. And the worst thing such an exposition can do is confuse the two.

In the course of this book I will also attempt to present a "fair and balanced" treatment of string theory (in a "non–Fox News" sense)–the source of most of the recent fascination with extra dimensions–and its offshoots. As we shall see, there are many fascinating theoretical reasons for physicists to be excited about working on these ideas. But that should not obscure the important fact that string theory has yet to demonstrate any definitive connection to the real world and, in fact, is a theory that thus far has primarily succeeded in generating more complex mathematics as time proceeds, any hype notwithstanding.

Because of the deeply ingrained nature of the concepts I want to deal with here, while science will form the core of our narrative thread, this book will present a broader history of ideas. This cultural context for the notion of extra dimensions is almost equally compelling, whether in literature or art. Science is not practiced in a vacuum, and, as I have argued, the very fact that the same ideas crop up, often centuries apart, may be telling us something, if not about the natural world, then at least about the human mind.

But what I ultimately find so striking about this story is a facet of science that mesmerizes me each time I visit a physics laboratory. While nothing may seem more esoteric than the notion of hidden extra dimensions, the scientific basis of all such theoretical speculations follows a sometimes circuitous path that however remains rooted in experiment. This remains true even if these experiments sometimes appear on the surface to be as far removed from these notions as baseball is from brain surgery.

Through this roundabout process, scientific progress has nevertheless been unmistakable. We fly in airplanes and launch rockets that explore the outer planets. We develop new medicines that extend our lives. We communicate electronically across the globe in an instant, sending messages that once would have taken weeks or months to arrive. Science is an arena of human affairs where we have every right to *demand* proof that new ideas work.

While Plato's beliefs about mathematics may seem distinctly modern, Greek philosophy as a whole was largely impotent in technologically empowering that civilization precisely because empiricism was missing from the equation. Natural philosophy had not yet evolved into science.

When it thus comes to the possibility that the three dimensions of space we experience are not all there is, I admit to being an agnostic. There are fascinating scientific and mathematical reasons to at least consider the possibility that our three-dimensional space is but the tip of a vast cosmic iceberg. At the same time, there is as of yet not a single compelling reason to believe that this is actually so.

By exploring the artistic, literary, and scientific bases of our current worldview, and taking the discussion up to the current threshold of our own understanding and our own ignorance, we will encounter some of the most fascinating developments of the human mind and some of the most remarkable discoveries about our own universe. Ultimately, I hope to provide you with a better perspective to help you decide on your own what seems plausible, and why. At this point, I believe it is anyone's guess.

As we embark on our tour, it may be worth quoting the cautionary advice of Antoine Lavoisier, one of the great scientists of the eighteenth century. Lavoisier was the father of much of modern chemistry but was executed during the French Revolution, which was itself based on an ill-founded notion of a "scientific" basis for human affairs. He is best known for his discovery of the profoundly important role of an invisible gas, oxygen, in the chemistry of our world. Regarding the emerging exotic science he helped found, Lavoisier warned: "It is with things that one can neither see nor feel that it is important to guard against flights of imagination."

CHAPTER 2

FROM FROGS' LEGS TO FIELDS

Why sir, there is every possibility that you will soon be able to tax it!
–Michael Faraday to Gladstone when asked
about the usefulness of electricity

The scientific realization that space and time might not be quite what they seem emerged from the unlikeliest of places: the nineteenth-century laboratory of a former bookbinder's apprentice turned chemist, then physicist, tucked away in the heart of London, over fifty years before Edward Abbott penned his mathematical romance of many dimensions.

Michael Faraday was a common man with an uncommon passion. In his lifetime he refused both a knighthood and the presidency of the Royal Society, preferring to remain, in his words, "just plain Michael Faraday." Perhaps his humble background forced him to develop an uncommon intuition about nature or at least an uncommon ability to develop pictorial explanations of natural phenomena that could bring otherwise lofty mathematical notions down to earth. Indeed, he claimed–no doubt sarcastically–to have written down a mathematical equation only once in his lifetime. Whatever its origin, he had an inherent predisposition against theoretical models that strayed even slightly beyond the constraints of experiment. It is thus ironic that Faraday ultimately provided the impetus for the creation of one of the greatest theoretical generalizations in the history of physics, a key that unlocked a door to a hidden universe. That key took a form no one could have anticipated in advance, and involved an act of serendipity in an experiment in a laboratory full of chemicals, wires, batteries, and magnets.

The experiment itself was disarmingly simple. A cloth-covered wire wrapped around one-half of a metal ring was connected to a switch connected to a battery. Another similar wire was wrapped around the other half, but hooked up to a device that could detect the flow of electric current through the wire. Since the two different wires were not in direct contact and the cloth wrapping insulated them from the metal ring, when the switch was closed–causing a current to flow in the first wire–there was no immediate reason to have expected a current to flow in the second. But to his amazement, Faraday discovered that at the precise instant that the first switch was closed, or opened again, and *only* in the instant when electric current either began or ceased to flow in the first wire, an electric current was mysteriously observed to flow in the second.

The uninitiated reader will at this point have at least two questions: (1) Why on earth did Faraday set up such a weird experiment in the first place? and (2) What has this got to do with space and time? The answers will require us to do some time traveling of our own.

Over half a century before Plato penned *The Republic,* the Greek playwright Euripedes had coined the name *magnets* for the odd lumps of ore found in the Greek province of Magnesia. The mysterious attraction of these objects to bits of iron fascinated the Greeks as it has fascinated generations of budding scientists in each of the twenty-six centuries since then.

The Greeks also discovered another invisible force, one between amber (when rubbed with fur) and bits of wood or fabric. This force did not receive its modern name for almost twenty centuries, however, until in 1600 the British scientist William Gilbert adapted the Greek word for amber, *electrum,* to its modern form, *electric,* to describe this strange attraction.

Following Gilbert's own studies, electricity and magnetism became the objects of intense scientific interest over the next two centuries. Electricity yielded to a simple mathematical description first, although it would take almost 170 years before the nature of electric forces between charged objects was fully described.

Red herrings, priority disputes, false leads, and theories without experimental basis all complicated the search for a fundamental understanding of these forces, as they sometimes do in science. The *Journal de Physique* wrote in 1781 words that seem disarmingly familiar in a current context: "Never have so many systems, so many new theories of the Universe, appeared as during the last few years."

One of the more colorful episodes in this saga involved two brilliant Italian scientists, Allesandro Volta and Luigi Galvani. The subject of the great debate between these two involved nothing less than frogs' legs. Galvani had discovered, in 1786, that electrical discharges could cause the leg of a dead frog to convulse. Ultimately, he was even able to make them convulse, simply by touching two different metallic plates to the frog's nerves. Galvani assumed that this metallic arc released some inherent electricity within the frog itself.

Meanwhile, Volta, who had developed sensitive instruments to detect the flow of electric charge, felt instead that somehow the electricity was produced by the contact of the two different metals. Ultimately, he was able to prove that this was in fact the cause of the dancing frogs, but more importantly, in the process he developed the electric battery, which introduced a valuable new tool for both science and technology.

In 1800 the American expatriate Count Rumford founded the Royal Institution in London and appointed the twenty-three-year-old chemist Humphrey Davy as its director. In the basement of this building Davy built a huge battery, based on Volta's principles, which he used to power a host of groundbreaking chemical experiments.

Davy was an imposing figure in British science, and his chemical experiments attracted the attention of scientists and laymen alike. One of these, a young bookbinder's apprentice, was fascinated with science and devoted his leisure time to its study. After attending a series of lectures given by Davy, Michael Faraday bound his carefully prepared notes in a volume and presented them to the great man, with a humble request to be considered for the position of Davy's laboratory assistant. In a lesson that many students have since learned–namely, it never hurts to flatter your teacher–Faraday was rewarded with the job of his dreams in that very year, 1813.

Meanwhile, on the Continent, strange new observations were underway that began to illuminate an intriguing hidden connection between the otherwise diverse phenomena of electricity and magnetism. It had long been suspected, given the various resemblances between electricity and magnetism (like charges repel, while opposite charges attract, just as two north or two south poles of magnets repel, while opposite poles attract, etc.), that perhaps these two forces were related in some fashion.

In the same year that Faraday joined the Royal Institution as Davy's assistant, Danish physicist-poet Hans Christian Oersted set out a challenge to himself and others to demonstrate that electricity and magnetism were indeed related. His quest was rewarded seven years later when he published a remarkable discovery: When an electric current flowed through a wire, it could change the orientation of a nearby compass. Oersted had discovered that electricity, when it flows, could generate magnetism.

It is difficult to describe the excitement that reverberated throughout Europe when Oersted announced his findings in a short paper, which was translated into various European languages from Latin within weeks. The day it was published in England, Davy brought it down to the laboratory and began working immediately to reproduce its results. Twenty-five years later Faraday reminisced about the repercussions of Oersted's work: "It burst open the gates of a domain in science, dark till then, and filled it with a flood of light." Once again, it's the image of moving from darkness to light.

The intense intellectual activity throughout Europe following the publication of Oersted's research was such that within several weeks the eminent French mathematician and physicist André-Marie Ampère developed a remarkable theory of how electricity could produce magnetism, which he later named electrodynamics. Based on a small amount of experimentation and a lot of guesswork and speculation, Ampère's original ideas were scattershot, but within a year or two they had came together to form the well-known theory that is quoted in physics textbooks today: Ampère reasoned and demonstrated that if currents running through wires could create magnets, and if magnets attracted or repelled, then two nearby wires with currents flowing in them should be repelled or attracted, depending upon the relative directions of the two currents.

One of the people whose critical examination helped Ampère ultimately refine his theory was the budding physicist Faraday. The year after Oersted made his discovery, in fact, Faraday published his own first significant discovery regarding electricity and magnetism. (Essentially all of his previous work had been on chemical analysis.) He discovered that small magnets would rotate around a wire with a current flowing through it, or alternatively, that a wire with a current flowing in it could be made to rotate about a fixed magnet. This established the peculiar nature of the magnetic force that was produced by moving electric charges, and ultimately verified key aspects of Ampère's ideas. The fact that the resulting force between the magnet and the wire was not merely attractive or repulsive, like the electric force between charges, but rather, pointed perpendicularly to an imaginary line joining the two objects (which would make one rotate around the other) was the first hint that the relationship between electricity and magnetism would require a completely new way of thinking. The simple, intuitive world that Newton unveiled with his brilliance was about to reveal its hidden underbelly.

It is interesting to note that in a letter written at the time to a friend in Geneva, Faraday talked about his early reticence in working on the subject associated with Ampère's "wild" theories:

> Theory makes up the great part of what M. Ampère has published, and theory in a great many points unsupported by experiments when they ought to have been adduced . . . [F]or myself, I had thought very little about it before your letter came, simply because, being naturally skeptical on philosophical theories, I thought there was a great want of experimental evidence.

Faraday went on to spend the next forty years of his life providing that evidence, and in what is perhaps one of the more profound ironies of physics, he ultimately provided the key theoretical idea that would reveal the true relationship between electricity and magnetism. While Oersted had shown that the former could create the latter, as early as 1822 Faraday wrote in his experimental notebook, where he recorded all his thoughts and ideas, the suggestion "convert magnetism into electricity."

Nine years later, on August 30, 1831, Faraday achieved the long-sought goal by means of his most famous experiment, described at the beginning of this chapter. But while Faraday demonstrated that magnetism *could* create electricity, it did so in a way that no one had suspected.

A normal magnet, no matter how strong, could not generate an electric current. However, a magnet whose strength *changed* could produce a current in a nearby wire. In his initial experiment Faraday created such a changing magnet simply by turning on and off a current in the first wire. As Oersted had already established, once a current was flowing in a wire, that wire acted like a magnet. Thus, during the brief period that the strength of the current rose from zero to its ultimate value, the corresponding strength of the magnet that it generated varied accordingly. It was only on the short interval surrounding the times that the circuit was either opened or closed in the first wire that Faraday noticed a current flowing in the second wire.

Faraday verified his idea that it was actually the changing strength of the magnet that caused a current to flow in the second wire by conducting a different experiment. Instead of turning a current on and off in the first wire, he simply moved a magnet closer and then farther away from the second. A current flowed as the magnet approached and again as it was withdrawn.

We now call Faraday's discovery induction, because one can induce currents to flow in wires exposed to magnets whose magnetic strength, relative to the wire, is changing. Faraday was justified in the promise he made Gladstone quoted in the epigraph to this chapter, because today we do tax this phenomenon, which has made possible most modern technology, as it allows us to produce electric power from sources such as falling water. If the water can be channeled through a tunnel, and made to spin a turbine holding several magnets within it, as the turbine spins around currents will be induced to flow in wires surrounding it. This is how we generate most of the electricity in the area of the United States where I currently live.

While Faraday's experimental discoveries therefore changed the face of modern society, they also changed our picture of nature. With his highly intuitive sense of nature, Faraday distrusted simple mathematical descriptions of phenomena, such as the force of attraction between two magnets. He preferred instead to formulate a physical "picture" of this force, so as a visual aid he suggested that throughout the space surrounding the magnets, one could imagine "lines of force." The direction of the force that would be experienced by another magnet that one might locate at any position would follow along the lines of force passing nearby. Similarly, the total number of field lines located near this point would signify the strength of the force. Ultimately Faraday used the same kind of visualization to describe the electric forces between charged particles, again without resort to mathematical equations.

Had Faraday been more comfortable with mathematics, he would have recognized that these "field lines" themselves had a simple mathematical description in terms that we now describe as a "magnetic field." A field is simply a function that assigns to each point in space some quantity. This quantity can be something as simple as a single number, or it can be something more complicated, such as a vector, which is a number plus a direction, appropriate to describe a force, for example.

The idea that magnets and charges might give rise to magnetic and electric fields, respectively, represented a major conceptual advance. From the time of Newton onward the question of how forces such as gravity actually act on distant objects had been a complete mystery. So-called instantaneous action at a distance seemed physically implausible–how did the earth know where the sun was in order to be attracted to it?–but a necessary, if unpleasant, fact of life. Faraday's fields solved this problem, at least in principle. If electric or magnetic fields exist throughout all of space, surrounding every charged object or magnet respectively (and for the moment one could ignore the question of how long it would take for such fields to develop around each such object), then a charged object or magnet located at a remote distance from another such object could experience a force due indirectly to that distant charge or magnet, but manifested directly via an interaction with the electric or magnetic field present in its own immediate vicinity. No direct action at a distance would be required!

Faraday reasoned that gravity, too, could be described in terms of lines of force, thus avoiding Newton's conundrum.

By the time that Faraday introduced these ideas in print, he was a well-established scientific figure, so his colleagues certainly took note of them. However, his descriptions were sufficiently vague that it is fair to say that most others were not convinced by them. For the case to become truly compelling it would require a physicist whose talents as a theorist were a match for those of Faraday as an experimentalist. Fortunately, such a theoretician had just moved to England at around the time Faraday was proposing his ideas.

The nineteenth century was full of towering mathematical geniuses, a number of whom pushed forward the frontiers of accepted knowledge, such as Newtonian mechanics. James Clerk Maxwell, however, in his short lifetime, left a legacy that is unmatched by any of them. He not only originated what is now the modern theory of gases, and the basis for the theory of statistical mechanics, which Boltzmann, Einstein, and Gibbs would later place firmly at the center of modern physics, but also completed the theoretical formulation of electromagnetism, the model prototype for the theories of all the known forces in nature. So complete and beautiful was his formulation that his equations for electrodynamics, now called "Maxwell's Equations," are emblazoned on the T-shirts of physics students and teachers throughout the world, who rely on them for much of what they do on a daily basis (the equations, not the T-shirts).

All these were conceived by a man who, before he died at the tender age of forty-eight, established the reputation of the Cavendish Laboratory at Cambridge, whose first director he was, as the major experimental physics laboratory in the world. Born and raised in Scotland, Maxwell did not have an auspicious youth. A private tutor who had been employed to teach him was not optimistic, reporting that he was a slow learner. Later Maxwell got the nickname "Dafty" from his schoolmates. By his teens he began to show mathematical promise, and studied at Edinburgh University and then Cambridge, where he ultimately received a fellowship. Nevertheless, he longed for his native Scotland and returned to Aberdeen to teach.

His treatment there, however, does not suggest that he gave any indication that he would eventually become known as perhaps the greatest theoretical physicist of the century. When Marischal College, where he was professor of natural philosophy, was merged with King's College to form Aberdeen University, two professorships were merged into one, and his post was given to the professor at King's, forcing Maxwell to seek another position. He applied for the professorship at Edinburgh University, which had become vacant, but it was given to one of his friends and former classmates instead. Maxwell was once again driven back down to England, where he accepted a post at King's College London, which he occupied until he was ultimately offered his position as Cavendish Professor at Cambridge.

While in London, Maxwell got to know Faraday, for whom he had immense respect. Both physicists thought in terms of physical pictures, although Maxwell's mathematical talent was sufficient to allow him to translate his ideas into precise mathematical formulations.

In 1856, while still in studying in Cambridge, Maxwell wrote a lengthy paper entitled "On Faraday's Lines of Force," in which he attempted to put Faraday's idea on a solid mathematical footing. This was the first step in his attempts to determine and formulate the laws of electrodynamics in a mathematically consistent fashion, which would culminate in his *Treatise on Electricity and Magnetism* (1873). By the time his work was completed, he had taken the geometric crutch of Faraday–the electric and magnetic lines of force, and the "fields" they represented–and turned them into entities as real as you or I.

As it was originally discovered, through the experiments of Oersted, Faraday, and their colleagues, the theory of electromagnetism was framed completely in terms of measurable physical entities (charges, currents, and magnets) and how they interact with one another. By trying to picture how these interactions operated, Faraday imagined space as full of electric and magnetic fields. Who would have guessed that the fields themselves could produce physical effects even if there were no charges, currents, or magnets nearby to respond to them? It would be disingenuous to say that the answer was as clear as the nose on your face, except that it is: The nose on your face *is* clear precisely because of these fields. It is these very fields that allow you to see.

Let's recap the rules of electromagnetism up until Maxwell. Oersted had discovered that currents (i.e., *moving* charges) could produce a force on magnets. Ampère had shown that these currents were in themselves magnets. Faraday discovered that changing the strength of a magnet put near a charge could produce a force on the charge.

What concerned Maxwell (as it had Faraday) was trying to find a unified understanding of these effects. What happened in the empty space between charges and magnets that could convey these forces? Both scientists, as they flailed about trying to understand the nature of the electromagnetic interaction, imagined this empty space as being filled with a remarkable amount of paraphernalia (invisible vortices, ball bearings, etc.) that might implement the action of Faraday's imaginary field lines.

Ultimately Maxwell realized that the magnetic and electric fields that Faraday envisaged throughout space might have a reality beyond their mere mathematical convenience, even if Maxwell himself probably still personally retained a physical picture of some "fluid" medium that permeated space, like the classical aether of Aristotle, with currents flowing within it.

But the mathematical discovery that Maxwell made that changed everything was simply the following: One could frame the laws of electromagnetism in terms of these electric and magnetic fields as fundamentals and not derived quantities. If moving charges would produce an ever-changing electric field and also a constant magnetic field, then perhaps the observation about currents and magnets could instead be framed as this: Changing electric fields can produce magnetic fields. And the observation about forces on charges being produced by moving magnets (which would produce changing magnetic fields) could be rephrased: Changing magnetic fields produce electric fields.

This subtle revision, with the fields taking center stage, could only truly have physical meaning if, in empty space, devoid of charges and currents, a measurable magnetic field could be produced purely by a changing electric field, and vice versa. Again, Maxwell led the way by showing that the mathematical description of electromagnetism was not consistent unless this phenomenon–occurring in empty space without physical changes and currents–could also occur, and he described precisely an experiment that would demonstrate just this effect.

But the biggest prediction–one of my favorite ones in all of physics–was yet to come. If I take a charge and move it, the electric field around it changes. That changing electric field in turn produces a changing magnetic field. But that changing magnetic field in turn produces a changing electric field. And so on, and so on, and before you know it an "electromagnetic disturbance" will propagate out into space. Maxwell could use the equations of electromagnetism he had derived to calculate the velocity of this disturbance in terms of two fundamental constants in nature: the strength of the electric force between charged particles, and the strength of the magnetic force between magnets.

When he did this calculation, he found that this disturbance would have the character of a wave, like a water wave, with crests and troughs not of water, but of the fields itself. Moreover, the speed that he calculated for this "electromagnetic wave" was familiar. It turned out to be the speed of light. This suggested, and it was later confirmed by experiments, that light itself might be waves of electromagnetic fields.

Maxwell's remarkable proposal–that light itself is an electromagnetic wave–occurred a full decade before Edwin Abbott wrote *Flatland,* and it would be over twenty more years before a young physicist working as a patent clerk in Switzerland would realize the full implication of this insight. Nevertheless, nature was competing with the literary imagination. Within less than seventy-five years of the discovery of the electromagnetic phenomena that power our modern civilization today, Faraday's imaginary crutches had become real, and they would ultimately force us to change the way we conceive of such fundamental concepts as space and time.

CHAPTER 3
THE ROAD TO RELATIVITY

We have no direct intuition about the equality of two time intervals. People who believe they have this intuition are the dupes of an illusion.
–Henri Poincaré, *La Mesure du Temps*

The eighth edition of the *Encyclopedia Britannica* appeared in 1878, just a year before James Clerk Maxwell's untimely demise. In that edition Maxwell penned an article entitled "Ether," in which he sardonically commented, "Space has been filled three or four times over with ethers." His critique was based on the fact that scientists had, over the years, proposed separate, distinguishable, but invisible media permeating all space, in which either light, heat, electricity, or magnetism might be conveyed. Maxwell felt that one of his great contributions, by demonstrating that light was an electromagnetic wave, was to reduce all of these separate "ethers" to a single medium, in which such waves might propagate.

Maxwell was so convinced that such a medium must exist that he actually set out to measure its effect on the propagation of light rays from the moons of Jupiter when the gas giant eclipses them, as seen from Earth, when our planet is moving at different speeds relative to Jupiter. In 1879 he wrote a letter acknowledging the receipt of data on Jupiter and its moons from the Nautical Almanac Office in Washington, D.C.

Maxwell reasoned that if one measured the apparent velocity of light at different times relative to Earth by measuring the time it took light to traverse the distance from Jupiter to Earth when Earth was moving in different directions in its orbit through the fixed ether in which the light rays presumably propagated, one could measure Earth's motion relative to this ether.

Whether Maxwell had sufficient time to adequately analyze the Nautical Almanac data before his death, or whether the data was good enough to even discern such a possible effect in principle, is now immaterial. The truth is, his proposal was doomed to fail, for reasons even he probably never imagined.

The first empirical evidence that the velocity of light did not obey the expected dependence on Earth's motion appeared less than two years after Maxwell's letter to Washington, in an experiment performed by the man who would eventually become America's first Nobel laureate in science, Albert A. Michelson. Michelson was on leave from the navy at the time, doing what all good would-be scientists living in the United States who wanted to get ahead then did–namely, spending time in the superior laboratories in Europe. In this case, he chose to work in Helmholtz's laboratory in Berlin.

Michelson, an experimental genius, had designed an apparatus that could detect a far smaller effect caused by the Earth's motion through the ether than Maxwell had proposed looking for. Instead of relying on data from observations of the Jovian system, Michelson could compare the round-trip travel time of two light rays traveling at the earth's surface in different directions with respect to the earth's motion around the sun–and thus also, presumably, with respect to the ether background. (Light rays traveling through the ether would presumably travel more slowly relative to the earth if they were battling an "aether headwind" as opposed to being propelled along by it, just as a golf ball hit into a headwind will travel more slowly, and hence cover less distance, than a ball hit into a tailwind. As a result, the round-trip travel time of a light ray should depend on its direction of motion relative to an ether headwind.)

Even though the predicted effect of the earth's motion through the ether was minute, Michelson's apparatus should have been able to discern it, but in 1881 he reported that his attempt to do so was unsuccessful. He was unequivocal in his conclusion: "The result of the hypothesis of a stationary aether is thus shown to be incorrect, and the necessary conclusion follows that the hypothesis is erroneous."

It is remarkable how willing Michelson was to throw out centuries of accepted wisdom on the basis of a single experiment, but while he was supremely confident in his results, the rest of the world was not. The eminent Dutch physicist Hendrik Lorentz, who was one of the few who seemed to even consider Michelson's data seriously, uncovered an error in Maxwell's theoretical analysis and thus distrusted the rest of the work. Both he and the eminent British physicist Lord Rayleigh urged Michelson to repeat the experiment with higher accuracy.

Thus it was that in 1887 Michelson, who had moved to Case School of Applied Science in Cleveland, teamed with chemist Edward Morley, from nearby Western Reserve University–a collaboration that presaged the merger eighty years later of Case and Western Reserve into my home institution, Case Western Reserve University–to perform one of the most celebrated experiments in modern physics.

The Michelson-Morley experiment definitively established that the velocity of light as measured on Earth was independent of a light ray's direction relative to the earth's motion around the sun. While Michelson jumped to the conclusion that this implied the ether did not exist (ultimately the correct conclusion), it is, in fact, not the only logical possibility. Rather, the results could have implied that for some reason the ether may have affected the measurement of light's velocity in ways that no one had yet understood.

Indeed, Lorentz's first question following the experiment, in a letter to Rayleigh, was whether there could be some error in the dynamic theory of electromagnetism that might explain the Michelson-Morley data. Lorentz continued to think deeply about this paradox, and in 1892 he argued that there was only one way he had come up with to reconcile their findings with commonsense notions about what should happen for observers moving with respect to each other. They would measure precisely the same round-trip travel time for light rays going in different directions with respect the earth's motion through a stationary ether if, somehow, lengths along the direction of motion with respect to the ether were foreshortened.

What Lorentz was in effect arguing was that the only way light rays would be measured to take the same time for round-trip travels independent of whether or not they were fighting an ether headwind would be if somehow lengths were also shortened along the direction of motion as the earth moved against any such headwinds. Since the distance the light ray traveled is calculated by its velocity multiplied by the time it travels, shortening the distance would cancel what would otherwise have been an extra travel time due to the slowing of light in these directions.

It was not so radical an idea to imagine that dynamic electromagnetic effects could cause lengths to so change. After all, if light is an electromagnetic phenomenon, and electric and magnetic forces are conveyed via the medium of the ether, then perhaps the electrically charged particles that make up the constituents of all atoms could be affected by their interactions with the ether as they pushed through it in a way that would move the atoms closer together. (In fact, the Irish physicist George Fitzgerald made precisely this argument in 1889, to derive precisely the same result, although it was unknown to Lorentz in 1892.)

Over the next twelve years Lorentz continued to think about the nature of electromagnetism in this context, and also about the mathematical properties of the theory that might determine what different observers moving with respect to each other would measure. In the process he made an observation that is implicit in Maxwell's equations but that had never been explicitly described. In 1895 he demonstrated that a moving charged particle would experience a force in a background magnetic field, because moving charges produce magnetic fields, and are therefore magnets and so must also experience forces due to other magnets.

I have always felt that it is precisely this revelation that carries the key to understanding why it was electromagnetism, and not some other force, that led Einstein to cause us to rethink our ideas of space, time, and motion. Ultimately, what the "Lorentz force," as it has become known, tells us is that what one observer measures as uniquely an electric force, another observer can measure as a magnetic force.

Think about it this way. If you are at rest with respect to some charged particle, and you observe it to move, you know it must have experienced a force, because things do not suddenly start moving without a force having acted on them. But the only force that a charged particle at rest can respond to is an electric force. Now, instead, imagine that you are moving at a constant velocity away from the charged particle. Relative to you the particle is moving backward, away from you. The laws of electromagnetism say that in your reference frame this moving charge must produce a magnetic field. If such a particle is then suddenly deflected in its path, you can measure this deflection and infer that the cause of this deflection was due to an external magnetic field acting on this current.

Thus, one person's electricity can be another person's magnetism. That is really the beauty of Maxwell's theory of electromagnetism. It demonstrated that electricity and magnetism are not only related, they are identical–merely different sides of the same coin. Different observers would measure the same phenomena, and ascribe them to either magnetic or electric effects, depending upon their state of motion. Since it is motion that relates electric and magnetic fields, it is perhaps not so surprising that light, an electromagnetic phenomenon, would cause us to rethink the nature of motion itself.

Albert Einstein was only five years old when the Michelson-Morley experiment was performed, but over the next eighteen years, while Lorentz, Fitzgerald, Rayleigh, and other well-known physicists were puzzling over the null results of Michelson and Morley, Einstein came to realize that the real problem was not reconciling Maxwell's theory with the Michelson-Morley finding (which he would later often claim not even to have known about at the time), but rather reconciling Maxwell's theory with the understanding of space and time that had prevailed in physics since the days of Galileo.

Again, with hindsight, the problem can be simply stated. One of Maxwell's greatest discoveries was that if light was an electromagnetic wave, one could calculate its speed from first principles, using solely quantities that could be measured in any laboratory associated with the strength of electric and magnetic forces.

But there is a fundamental, hidden problem with this result. It had long been recognized–indeed, since the time of Galileo and later Newton–that the laws of motion as measured by an observer moving at a constant velocity (say, a person on a train or plane) are the same as for an observer standing still. Think about throwing a ball in the air or playing catch. If you are on a plane or train that is moving in a straight line, and you throw a ball up in the air, you will see exactly the same thing that you would see if you threw the ball while standing still. This is to say, you won't feel as if you are moving. If the windows are covered, and there are no bumps, and the engines are not making any noise, there is, in fact, really no way to know if you are moving or standing still.

Galileo first recognized this fact about motion and codified it, stating that there is no way to distinguish between observers at rest and observers in constant motion. That principle is literally the foundation on which all of our understanding of modern physics was based. We now call this "Galilean relativity."

However, as Einstein realized from his teenage years onward, there is a problem reconciling Galilean relativity with Maxwell's discovery about light. For, if the speed of light can be calculated from fundamental constants that can be measured in a laboratory, and if observers in laboratories moving at a constant velocity with respect to each other should observe the same results as observers in laboratories at rest, then this would imply something remarkable. Since all such observers should measure the same fundamental constants of nature, in terms of which they could each calculate the speed of light rays that they would measure in their laboratories, then all observers, regardless of their state of motion with respect to an ether background, should measure the same speed of light.

This result is, of course, precisely what the Michelson-Morley experiment seemed to demonstrate, but it also leads to a paradox if light is a wave in an ether. It is like saying that, if you are driving a car along a river, the waves moving in the water would appear to move along relative to you at the same speed that they would be measured to move relative to someone sitting on the shore. That is silly, because if your car is moving along at the same speed as the waves, they will be stationary with respect to you, but not to an observer on the shore.

This is so counterintuitive that it perhaps explains why the best physics minds in the world spent much of the two decades after the

Michelson-Morley experiment trying to find a way to dynamically change the predictions of Maxwell's theory in different ways to accommodate it, rather than accepting that the theory in fact required this result. Einstein, on the other hand, accepted this implication of Maxwell's theory at face value, because the theory perfectly described all other measured aspects of electromagnetism. Instead, he recognized that to accommodate it one would have to revise other aspects of our understanding of the world.

The first person to suggest that one must begin to think along these lines was not Albert Einstein, but the famous French mathematical physicist, Jules Henri Poincaré. A leading scientific intellect who had a philosophical bent as well, Poincaré realized as early as 1898 that we might have to alter basic notions regarding the objective meaning of various concepts of space and time to account for the fact that the occurrence of events at distant points could only be relayed to us after a finite time. It was in this context that he uttered the words quoted at the beginning of this chapter.

Poincaré even discovered in 1905, the same year that Einstein published his first paper on special relativity, that the equations of electromagnetism remained unchanged if measurements of space and time change for different observers in relative motion in precisely such a way as to reproduce the "Lorentz contraction"–as he then referred to it–which Lorentz had earlier proposed to reconcile the negative result of the Michelson-Morley experiment. Poincaré even demonstrated that the different observers who synchronize their clocks by light signals may have different notions of simultaneity.

It is remarkable that in spite of discovering all of these pieces, Poincaré never fully put the puzzle together. He remained committed both to the ether and to a dynamic origin for the contraction of bodies along their direction of motion relative to the ether.

It remained for Einstein to demonstrate that Maxwell's equations, when combined with the ideas of Galilean relativity, provided all that was necessary to resolve the paradoxes of electrodynamics without additional dynamic hypotheses. All that one had to do was dispense with the absolute definitions of space and time, and give up the notion of an ether at absolute rest in the universe.

This was no small intellectual leap. But even Einstein did not anticipate at the time that this step would literally add a whole new dimension to the universe.

CHAPTER 4
THE FOURTH DIMENSION

Henceforth space by itself, and time by itself, are doomed to fade away into mere shadows, and only a kind of union of the two will preserve an independent reality.

–Hermann Minkowski

Physicists at the turn of the twentieth century were understandably reluctant to abandon the security of a sensible worldview that up to that point had appeared to successfully describe the universe. But ultimately, once the hidden connections that underlie electromagnetism came into clear focus, there was no turning back, and the road that began at "Let there be light" led straight into a fourth dimension.

First, the sensible worldview: If I am running away from you, and someone behind you throws a ball at you, I would expect that the ball would appear to be traveling faster relative to you than it would to me. Common sense similarly suggests that two different observers in relative motion will measure the same light ray to travel at different speeds relative to each of them in, say, one second.

Now the problem: Electromagnetism only makes sense, in a world where all laboratories measure the same strength of magnetism and electricity, if the light ray mentioned above approaches each observer by the same distance in the same time, even if the observers are moving apart.

Now for Einstein's solution: If the light ray is to be so measured, then each observer must use different measures of distance or time. Upon careful analysis Einstein determined that both measurements must differ. Specifically, Einstein demonstrated the following implications of the strange behavior of light, in what we now call Einstein's special theory of relativity:

(i) Remote events that are simultaneous to one observer will not be simultaneous to another observer moving with respect to the first.
(ii) Clocks carried by an observer moving with respect to another observer will be measured by the latter to be running slowly.
(iii) Objects carried by an observer moving with respect to another observer will be measured by the latter to be foreshortened along the direction of their motion.

Einstein arrived at each of these bizarre conclusions by doing what he called *gedanken,* or "thought experiments," that get around the fact that on human scales our perceptions of space and time are vastly different from what they would be if we could travel at near light-speed. In this, he followed the spirit of Poincaré's thinking. As Poincaré first pointed out, our knowledge of remote events is always indirect, because remote events are, after all, remote. We may feel like something we see happening across the room is happening at the same moment as we see it, so that we are a "part" of the event, but that is merely an illusion brought about by the incredibly fast speed of light.

Consider a class photograph. We are accustomed to thinking that it reflects a single frozen instant in time, when all of the bright young faces are captured as an enduring memory. But, strictly speaking, this is not accurate. Just as the different rows of students are spread out in space, the photograph reflects an image that is also spread out in time. The light reflected from the faces of the children in the back row arrives at the camera lens at the same instant as the light from the faces of the children in the front row only if the light from the back row began its voyage slightly earlier. The time difference is imperceptible, perhaps a billionth of a second or so, but it *isn't* zero. If each row were separated from the row in front by, say, a hundred million miles, instead of a few feet, then the students in the back row could easily have left their seats by the time the students in the front row had begun to pose for one and the same photograph. This is because the light from the back row would take about ten minutes to reach the front row, and would thus reach the camera at the same time as light emitted from the front row ten minutes later.

In an astronomical context, this is always true. When we lock up at the sky at night, the images of the individual stars reflect moments spread out by hundreds if not thousands of years.

We are accustomed to this phenomenon in a reverse context because of the fact that sound travels much more slowly than light. When we see lightning strike in the distance, and we hear the thunder clap many seconds later, we know that they relate to one and the same event, even though we experience its different aspects at different times. It is equally true however, that things we experience in a single instant can reflect not one event, but many separate ones.

Einstein imagined a scenario where this would be explicit. Picture, for example, a train so long that light from one end of it would take several seconds to reach the other. Now picture that you are in the middle of the train. Now picture, finally, another implausible series of events: Lightning strikes both ends of the train at exactly the same instant.

How do you know that the two lightning bolts hit either end of the train at the same time? Simple: You see the two flashes in your car at the same instant. Since you are in the middle of the train, you know that, even accounting for the fact that it has taken some time for the images to reach you, since the time for both images to reach you is the same, the flashes must have been simultaneous.

Now, what about someone on the ground whom you see directly opposite you at exactly the instant when the lightning bolts struck the ends of the train (not later, when you actually see the flashes!)? What would she see (assuming the flashes were bright enough for her to see them as well)? Well, since you are moving with respect to her, by the time you see the flashes she must now be closer to one end of the train than the other.

Thus, the light from one of the flashes must have passed her location before it made its way to you. Hence, she will see one of the flashes before the other. But since she was opposite you when the lightning hit either end of the train, and was thus also midway between the flashes, and since she sees one before the other, she must infer that one of the flashes hit before the other.

What is wrong with this picture? Well, in a sensible universe the person on the ground would indeed see one flash before the other, and the person on the train would see both flashes at the same time. But the person on the train, whom the person on the ground would see moving toward one flash and away from the other, would also be able to (if she had the proper apparatus) measure that the light ray from the side of the train that she was moving toward would be traveling relative to her faster than the other light ray, which she would be moving away from. Thus, although she saw both flashes at the same time, she would indeed be able to infer from her measurements that one event had to have occurred before the other in order for her to experience them simultaneously, in agreement with the assessment of the person on the ground.

But the universe isn't so sensible. Maxwell, Einstein, and experiment all tell us that both observers will measure the speed of both light rays relative to themselves to be exactly the same, and, as a result, each observer is forced to a different conclusion about the simultaneity of the two events.

It is important here not to think that one observer is right, and one is wrong. They are both right. There is not a single experiment either person can do to change her own perception of the events or to prove the other person wrong. If they could, then one of them would be able to prove that she was at rest while the other person was moving. But that is the whole point. There is no absolute rest frame with respect to the speed of light. All observers are equivalent.

So that means that whether or not distant events are simultaneous depends upon who is doing the observing. There is no absolute "now." "Now" means something unambiguous only right where you are. Anything you conclude about "now" elsewhere is simply an inference, and it is unique to you. To put another way, "now" is relative.

It is also important not to think that any sense of "now" is therefore completely arbitrary. It is just as constrained after Einstein's *gedanken* experiment as it was before. Each observer can base a consistent reality on what she sees, and she can count on the fact that events never precede causes, and so on, even if it turns out that for one observer one event may happen before another, while for another observer precisely the opposite may be true. It turns out that the mathematics of relativity happily only allow this reversal in temporal ordering for events witnessed by different observers whenever the events are sufficiently remote in space and close in time, so that one event cannot have been the cause of the other. Put another way, if a signal can travel between the events in the time between them, then all observers will end up agreeing about which happened first, even if the observers might disagree about how much time had elapsed between them.

But just in case you were beginning to think things might be sensible after all, consider the following: The same type of reasoning that led Einstein to recognize that simultaneity was relative led him to recognize that measures of length and time themselves were also relative.

For example, let us return to our train example. When the lightning struck simultaneously (for the observer on the train), let us say it scorched the tracks at the same time. Thus, that observer can come back later and measure the distance between the scorch marks to determine the length of the train. But the observer who was on the ground at the time will call foul. She will insist that because one lightning bolt hit before the other, and during the time between the two events, the train was moving, that, the scorch marks on the ground represent a distance that is *longer* than the actual size of the moving train. In short, the observer on the ground who sees the train moving past will insist that the train is *shorter* than will an observer on the train, who is at rest with respect to it.

So far so good. Moving objects are measured to be contracted along their direction of motion. In fact, this contraction is precisely that calculated earlier by Lorentz (which was dubbed the "Lorentz contraction" by Poincaré) when he tried to make sense of the Michelson-Morley experiment. But here the resemblance ends. In Lorentz's worldview, where there was an ether and a universal rest frame, moving objects could be contracted relative to those standing still. But in Einstein's universe, which happens to be the one we live in, all motion is relative. There is no universal rest frame and no ether.

So, for a person on the moving train, it is the person on the ground who can be said to be moving past, in the opposite direction. And exactly the same type of reasoning as given above will convince you that the person on the train will measure the lengths of objects at rest with respect to the person on the ground to be shorter than will the person on the ground! Thus, each observer will measure the length of objects at rest in the other person's frame of reference to be shorter. The Lorentz contraction is not absolute; it is relative.

Once again, the relative nature of the Lorentz contraction should not lull you into assuming that it is not real. It is as real as the nose on my face, whose size will, of course, depend upon who is viewing it. This is illustrated by my very favorite paradox from relativity. Thankfully for you, it is the last one I will attack your brain with here.

Say I have a fast sports car–a *really* fast one, which can travel at a large fraction of the speed of light, where the mysterious effects of relativity become more apparent. After all, if you consider the *gedanken* experiments I have discussed above, clearly the discrepancies about length and time between observers are related to how far the train could have traveled during the time the light rays crossed it. To have observable effects, one needs either very large trains or very fast ones.

Well, say I am moving past you at a very large fraction of the speed of light. My car will therefore be measured by you to be shorter than I will measure it to be. Now, say you have a garage with two doors, one at either end, into which I am driving. If my car is ten feet long to me, say it would be measured to be six feet long by you. Say your garage is eight feet in depth. Then, for you it should certainly be possible to quickly close the front door of your garage after my car has entered and continues speeding along, completely enclosing it within the garage. You would then hopefully run very quickly to open the door at the rear of the garage so that my speeding car would not run into it.

Relativity tells us that this is certainly possible, at least in principle. But now there is a problem. In my reference frame, it is your moving garage that is shorter. To me, appears to be only five feet long, and there is no way that my car will fit within it!

Am I doomed to crash? Well, if I do hit a door, both observers would have to agree that such an event happened. (After all, they can come back together afterward and see the tangled mess, if both people are still alive.) So, if one observer sees me making it through the garage safely, then I must have done so. Rather, I will insist that my car and I were never entirely within the garage, because I will measure the order in time of the remote events, including opening and closing the garage doors, to be different than will the observer on the ground. I will insist that, for example, the rear door of the garage was opened before its front door was closed. Thus, as I sped through, the front of my car exited the back of the garage before its rear end passed through the front of the garage.

The point is that each observer's reality is real. For you, my car was completely inside the garage. For me, it never was. There is no experiment you can perform that will prove me wrong, and vice versa.

At the same time, it is clear that the contraction, while real, is still very much in the eye of the beholder. Or, as Einstein would say, measurements of length are relative.

A similar relativity occurs for the slowing of clocks. If I am moving very fast relative to you, you will measure my clocks to be running slowly. I will appear to you to age more slowly if you watch me recede into the distance. But I will in turn measure your clocks to be running slowly as, well, you will appear to me to age more slowly.

At this point a conventional reaction to the implications of relativity is to throw up one's hands and decide that the world has no order in it whatsoever and that there exist no absolutes. Everything is relative, so anything goes! Indeed, this was the reaction of many artists and writers in the early part of the twentieth century to the results of relativity, as I shall soon discuss. But even if it feels justified, this is not the correct response.

Hermann Minkowski had been one of Einstein's mathematics teachers in Zurich–in fact, one of the few whose lectures Einstein actually enjoyed. In 1902 Minkowski moved to the University of Göttingen, where one of the most renowned mathematicians of his time, David Hilbert, was located. Interestingly, Hilbert would later help Einstein provide the mathematical tools that would change our picture space and time in profoundly new ways. But well before that, the first Göttingen mathematician to have had such an impact was Minkowski.

In 1908 in Cologne Minkowski gave a lecture entitled "Space and Time," which created a tremendous stir and has since been recognized as a watershed moment in our understanding of physical reality. The epigraph of this chapter is from that lecture, which began with words that are both enticing and particularly significant for a mathematician to have uttered: "The views of space and time which I wish to lay before you have sprung from the soil of experimental physics, and therein lies their strength. They are radical."

In his speech, and in the more technical paper that accompanied it, Minkowski delivered exactly what he had promised. By the time he was finished, space and time could no longer sensibly be individually discussed, and only a union of the two, which we now call space-time, was understood to retain any independent reality.

The seeds of Minkowski's realization lie in the example I presented involving Einstein's long train. Recall that simultaneous lightning bolts for an observer on a moving train provide an ideal method for her to measure the length of the train. She merely has to later disembark, return to the scene of the lightning strikes, and measure the distance between the scorch marks on the tracks.

Now, also recall that an observer on the ground will contest this measurement, arguing that the two lightning bolts were not simultaneous and therefore the scorch marks represent events that happened at two different times at either end of a moving train. Thus, the distance between the scorch marks must represent a larger distance than the true length of the train.

Let us then consider what this implies by thinking in terms of what, precisely, is meant by a measurement. The observer on the train measures an interval in space. That is, after all, what a measurement of distance is. For the observer on the ground, however, this same measurement involves an interval in space *and* time.

Seen from this perspective, perhaps it is not surprising that the individual distance and time difference measurements for the two observers differ. To visualize this a little more dramatically, let us imagine two observers in Plato's cave. One of them sees the following shadow on the cave wall, in the morning:

Later in the day, the other observer sees this one:

Has the person whose shadow they have seen at different times of day changed in height? No, of course not. Rather, the sun is higher in the sky, and the length of the shadow on the back wall of the cave will change accordingly.

Let's simplify the issue. Imagine the cave dwellers are viewing the shadow of a transparent ruler:

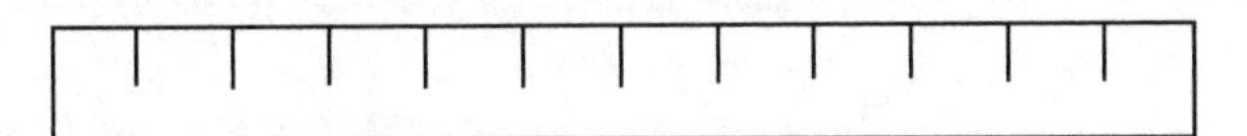

Now suddenly the shadow changes:

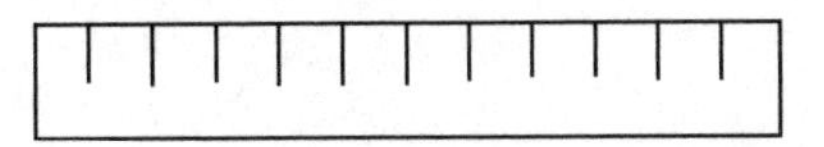

The shadow-ruler has inexplicably changed in length. How was this possible? Simple: The original ruler was rotated with respect to the light source. As seen from above, the two different situations appear as follows:

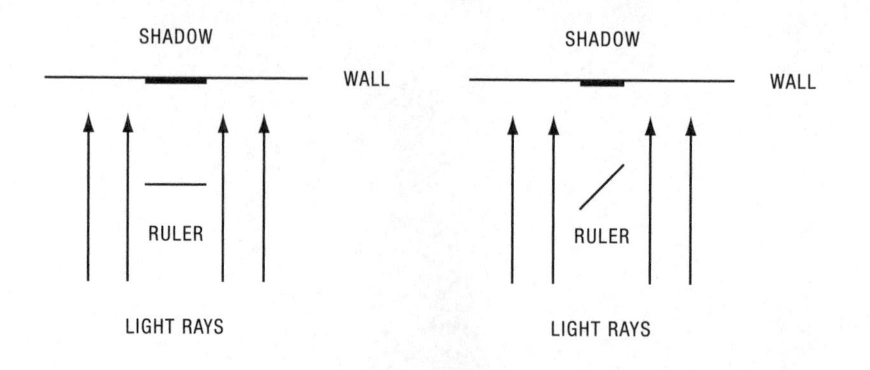

The length of the original ruler has certainly not changed by this rotation, but the *projection* of this three-dimensional object onto the two-dimensional wall at the back of the cave has. Physicists in this cave-dwelling society may initially be baffled by the fact that the lengths of shadow-objects are apparently not absolute. But eventually someone would intuit that the objects being observed behave as two-dimensional projections of three-dimensional objects that can be rotated perpendicular to the wall. Mathematically, there is a quantity that is absolute and doesn't change under such rotations–namely, the length of the original ruler. If this ruler has a length *L,* while the length of the shadow-ruler (i.e., the projection of *L* on the cave wall) is *X,* then a cave mathematician, who, for the sake of argument we might call Pythagoras, might suggest that there is a quantity, *L,* whose value does not change, and that is given by the relation $L^2 = X^2 + Y^2$, where *X* is the projection of the ruler on the cave wall and *Y* is the projection of *L* perpendicular to the cave wall:

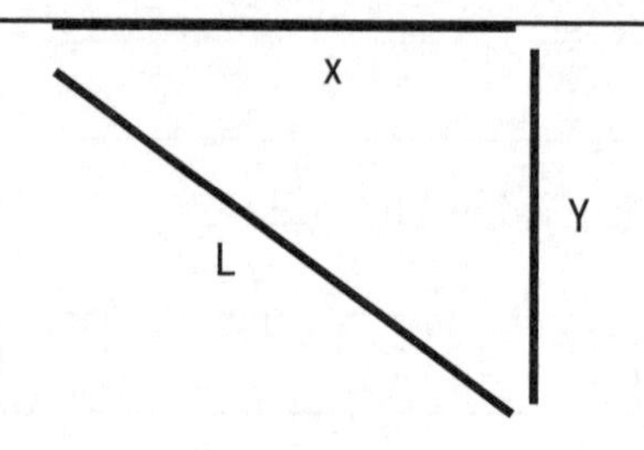

By now you don't have to be Einstein to see where we are heading. What Hermann Minkowski realized is that there is a similarity (but *just* a similarity) between this scenario and what occurs, according to relativity, for observers in relative motion measuring the same object.

Recall that the speed of light in empty space, c, is measured to be the same by all observers. Say one observer measures the distance traveled by a light ray in some time t to have a value d. Since distance traveled is determined by the speed of the light ray times the time it travels, this observer thus finds $d=ct$. Any other observer moving with respect to this observer may in general measure a different length d' and time t', but they must find $d'=ct'$ if they are to determine the same speed relative to them for this light ray.

Thus, at least for a light ray, different observers in relative motion will measure distances and times such that the combination $d^2-c^2t^2=d'^2-c^2t'^2=0$ for any light ray. While this is manifestly true for a light ray, it turns out that this combination will be measured to be the same by all observers for any two "space-time" events measured to be separated by a distance d and time t for any one of them, so that $d^2-c^2t^2=d'^2-c^2t'^2$ for, all events separated in space and time even if the combination is not zero (i.e., the two distances and times are not for points connected by the trajectory of a light ray). This will be true even though the separate observers will in general arrive at different separate measurements of d and t.

Minkowski realized that this particular combination of distance and time, which Einstein recognized remains invariant between observers in relative motion, is strikingly analogous to the way the different length projections of a ruler can be combined to always produce the same value–namely the length of the ruler itself–regardless of its orientation. Except for the weird minus sign (i.e., $d^2-c^2t^2$ instead of $d^2+c^2t^2$), which we will discuss shortly, the combination is the same.

Thus, the exotic results of Einstein's relativity can be understood by analogy to the two-dimensional cave example. In the latter case, different observations of the same object appeared inconsistent because each presented a different two-dimensional projection of the same three-dimensional object. In our universe, different observers in relative motion are simply presented with different *three-dimensional* slices of an underlying *four-dimensional universe* where space and time are tied together.

Minkowski called the mathematical combination $d^2 - c^2t^2$ the "space-time distance" between the events, to distinguish it from the three-dimensional, purely spatial distances we are used to. Just as rotations in regular space can change projections, so, too, can relative motion change the separate time and space intervals measured by different observers, while the space-time distance is preserved. Indeed, motion reproduces certain aspects that are reminiscent of rotations. As Einstein's train example makes clear, one man's space interval can be another man's time interval.

With this unveiling of what we now call "Minkowski space," Minkowski delivered on the promise of his Cologne lecture. Our Plato's cave illustration merely makes literal his metaphorical exclamation that heretofore space by itself and time by itself would fade away into mere shadows.

From 1908 onward, three-dimensional space and the seemingly distinct and unrelated one-dimensional progression of time became inextricably linked together. What had begun with tentative inklings in basement laboratories filled with compasses and currents had blossomed into a whole different perspective of our universe to be explored and understood.

This four-dimensional space that we discovered we occupy, however, differs dramatically from the world that Edward Abbott envisaged in the plaintive pleas of his *Flatland* hero. The weird relative minus sign between the spatial part and the time part of space-time distance (remember that for normal spatial separations, the square of total distance between two points is the *sum* of the squares of the individual projections, with no minus signs) changes everything, so that and space are tied together in a way that is quite unlike the way *up* and *sideways* are tied together. We cannot walk into time as we can apparently walk into space, nor, as far as we yet know, can we back up. Time travel is so exotic compared to motion in space that entire movies and (fictional) books have been written to consider this possibility. The minus sign fundamentally seems to distinguish between space-time intervals that are "timelike" compared to those that are "spacelike." (Minkowski himself coined this terminology.)

Physics was thus left at the brink. A fourth dimension had been discovered, but not the one that Abbott had imagined. But people most often hear what they want to hear, and consequently they often tend to interpret the new results of science in terms that justify their previous expectations. Thus, the feature that makes Minkowski space special, while profound, was overshadowed by the newfound freedom of action offered by Einstein's special relativity, and the promise of a "fourth dimension."

But Einstein was not yet finished with space and time.

CHAPTER 5
DISTURBING THE UNIVERSE

What is derived from experience has only comparative universality, namely, that which is obtained through induction. We should therefore only be able to say that, so far as hitherto observed, no space has been found which has more than three dimensions.

–Immanuel Kant, *Critique of Pure Reason*

For Kant, space existed in the mind, as a backdrop for all of our experience. From his perspective Euclid's fundamental axioms of geometry were a priori necessary features of a universe in which thinking beings could live. Kant felt that these axioms were not derived from experience or experiment, for if they were, they would merely be provisional, not absolute.

Well, Kant was correct in at least one respect: The postulates of Euclid, in particular his famous fifth postulate–that there is only one line that can be drawn through any point that does not intersect with (i.e., is parallel with) a given line–cannot be derived from fundamental principles or from experience. That is not the case, however, because they are intrinsic to our existence. It is, rather, because they are not universally true. On a sphere, for example, lines of longitude are parallel, but they all meet at the North and South poles.

Such, it seems to me, is the limitation of much of philosophy: It is often subsumed as empirical knowledge supplants pure thought. The irony in this statement is that Einstein's most significant contribution to human knowledge comes as close as any major development I know of in the history of physics to something akin to pure thought. I refer to Einstein's general theory of relativity, which he developed in the decade after his formulation of special relativity in 1905. The term *general* here refers to the fact that special relativity applied to observers in constant relative motion. What general relativity did was to extend these considerations to accelerating observers. Remarkably, in the process, it turned out to be a new theory of gravity!

That is not to say, however, that Einstein's general relativity was motivated by mathematical concerns alone, either the beauty of tensor algebra, which made his theory calculable, or that of Riemannian geometry, which Einstein had to master in order to ultimately describe curved space. Far from it. The origins of Einstein's general theory of relativity stem from the same type of thought experiments involving physical phenomena that led to the special theory. In this case they came about as Einstein was pondering Newton's law of gravity, electromagnetism, and special relativity in 1907, a year in which he later stated he had had "the happiest thought of his life."

We have already seen how the relationship between electricity and magnetism implies that what one observer measures as a magnetic force, another could measure as an electric force. This "observer- dependence" of electromagnetism played a key role in the development of special relativity and the unification of space and time into space-time. Perhaps not surprisingly, a similar notion played a central role in Einstein's thinking when, in 1907, while considering Newton's gravity, he suddenly realized that it, too, was observer dependent.

He reasoned as follows: An observer who is free-falling in a gravitational field–like someone who jumps out of a plane–feels no gravitational forces at all. For this observer, the gravitational field is undetectable (at least until the rude awakening, followed by a quick demise, upon later hitting the ground). Ignoring any effects of air resistance, an object "dropped" from such an observer's hand would fall at the same rate of acceleration as that of the observer, so it would remain at rest relative to the observer. For all intents and purposes, gravity wouldn't exist for this individual. In this regard, as would be equally true for Galileo's observer moving at a constant speed in the absence of gravity, such a free-falling observer would have every right to consider herself at rest, because all objects at rest in her frame would remain at rest if no other (nongravitational) force was applied to them.

In this sense gravity, like electricity or magnetism, seems to exist truly in the eye of the beholder. But this picture is true only if all objects fall at the same rate. If a single object accelerated at a different rate from all other objects in a gravitational field, the whole notion that gravity might be invisible would fall apart. A free-falling observer would see this object as accelerating relative to her, and thus would be able to conclude that some external force was acting upon it.

This idea–all objects fall at the same rate due to gravity, independent of their composition–Einstein labeled the Equivalence Principle, and it was central to his development of general relativity. Only if it remained true could gravity arise as an accident of ones circumstances, just as the electric force that one might experience could actually be due to a distant, changing magnetic field.

While a violation of the equivalence principle would put an end to any chance of "replacing" gravity with something more fundamental and less observer-dependant, it is not obvious from this example what one might actually replace it with. Once again, Einstein provided a thought experiment that showed the way.

If falling in a gravitational field can get rid of any observable effects of gravity, accelerating in the absence of one can create the appearance of a gravitational field. Consider the following famous example. Say, for some inexplicable reason, you are in an elevator deep in space. As everyone who has ever been in an elevator has experienced, when it first starts to accelerate upward, you feel slightly heavier; namely, you feel a greater force exerted by the floor on your feet. If you were in outer space, where you would otherwise feel weightless, and the elevator you were in started to accelerate upward, you would feel a similar force pushing you down against the floor.

Einstein reasoned that, if the equivalence principle was indeed true, then there is no experiment you could perform in the elevator that could distinguish between whether that elevator was accelerating upward in the absence of a gravitational field, or whether it was at rest in a gravitational field, where the force the observer would feel pushing her down against the floor would be due to gravity.

So far so good. Now, imagine what would happen if the observer in the accelerating elevator were to shine a laser beam from one side of the elevator to the other. Since, during the time the light beam was crossing the elevator, the elevator's upward speed would have increased, this would mean that the light ray, which is traveling in a straight line relative to an observer at rest *outside* the elevator, would end up hitting the far side of the elevator somewhat below the height where it was emitted, relative to the floor of the elevator.

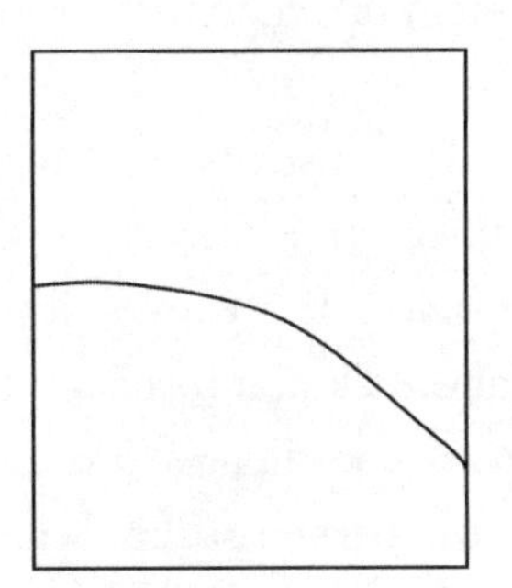

Now, if gravity is to produce effects that are completely equivalent to those we would measure in an accelerating system this would mean that if I shined a laser beam in an elevator at rest in a gravitational field (say, on Earth), I would also see the light ray's trajectory bend downward. (Of course, the effect would be very small, but since we are doing a thought experiment here, we are free to imagine an arbitrarily accurate measuring device.)

But, special relativity tells us that light rays move at constant speed in straight lines. How can we reconcile this behavior with what you would measure in the elevator? Well, one way to go in a straight line and also travel in a curve is to travel on a straight line on a curved surface.

This realization led Einstein on a long mental journey in the course of which he was drawn to the inescapable conclusion that space and time are not only coupled together, but are also themselves dramatically different than we perceive them to be. Space, and to some extent time, can be curved in the presence of mass or energy. The result was perhaps the most dramatic reformulation of our understanding of the underlying nature of the physical universe in the history of science.

Einstein's journey was replete with false starts and dead ends, and the slowly dawning acceptance that mathematical concepts that he had vaguely been exposed to while a student might actually be useful for understanding the nature of gravity. In 1912 Einstein finally realized that the mathematics of Gauss, and then Riemann, which described the geometry of curved surfaces and ultimately curved spaces, held the key to unlocking the puzzle he had been wrestling with all those years. By November 1915, after almost having been scooped by the best mathematician of that generation, David Hilbert, Einstein unveiled the final form of his "gravitational field equations."

Einstein's equations, as we usually call them, provide a relation between the energy and momentum of objects moving within space and the possible curvature of that space. There are at least two fascinating and unexpected facets of this relation. First, it turns out to be completely independent of whatever system of coordinates one might use to describe the position of objects within the curved space. Second, and true to the spirit of special relativity–which by tying together space and time also turned out to tie together mass and energy–energy becomes the source of gravity. In general relativity, however, such energy influences the very geometry of space itself–a fact that makes general relativity almost infinitely more complex and fascinating than Newton's earlier law of gravitation. This is because the energy associated with a gravitational field, and hence with the curvature of space, in turn affects that curvature.

In the jargon of mathematicians, general relativity is a "nonlinear" theory. While technically speaking this means that it is difficult to solve the relevant equations, in physical terms it means that the distribution of mass and energy in space determines the strength of the gravitational field at any point, which in turn determines the curvature of space at any point, which in turn determines subsequent distribution of masses and energy, which in turn determines the curvature of space, and so on.

Nevertheless, in spite of the difficulty of dealing with these equations, the single fact that affected Einstein during that fateful November in 1915 more deeply than perhaps any other discovery he had made in his lifetime was the realization that the mathematical theory he had just proposed explained an obscure but mysterious astronomical observation about the orbit of Mercury around the sun.

One of the most successful and stunning predictions of Newton's law of gravity is that the orbit of planets around a central body such as the Sun should be described by mathematical curves called ellipses. That the planetary motions were not perfect circles had first been discovered, somewhat to his dismay, by Johannes Kepler, and in short order Newton proved that his universal law universally implied elliptical orbits.

Nevertheless, in 1859 the French astronomer Urbain Jean Joseph Le Verrier discovered that the orbit of Mercury was anomalous. Instead of returning exactly to its initial position after each orbit, the planet advanced slightly, so that rather than forming perfect ellipses, the orbits traced a figure that was more like a spiral, with the axis of each successive orbit being slightly shifted compared to the one before it, as shown in an exaggerated view below:

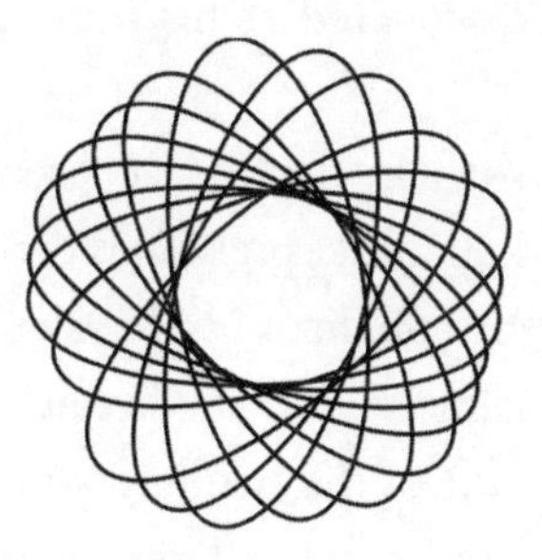

This "precession" was extremely small, measuring only about 1/100 of a degree per century. Nevertheless, in physics, as in horseshoes, being merely close is not good enough; if Newton was correct, there should be no such precession. Barring the presence of some new, undiscovered massive body nearby exerting a gravitational pull on Mercury, the only way such a precession could be explained was to slightly alter the nature of Newtonian gravity.

Beyond its profound underlying physical implications, this is precisely what Einstein's general theory does, making a small correction to Newton's law. It turns out that when the force law is no longer precisely as Newton described it, then a precession is predicted. Einstein, to his credit, was able to derive an approximate solution to his equations that was accurate enough to predict the precession of Mercury's orbit, and to his immense surprise and satisfaction, the prediction was precisely in agreement with this half-century-old puzzling result.

Years afterward Einstein recalled that, upon discovering this agreement between prediction and observation, he had the feeling that something had actually snapped within him. He suddenly realized that his journey of the mind had led him to more than mathematical fantasies. He said he was so excited that he had palpitations of the heart.

Later in his career, Einstein would become more enamored with the simplicity of the mathematical principles that were the foundation for general relativity. But I think it is crucially important to recognize–and I shall have cause to return to this theme–that what distinguished Einstein the physicist from Hilbert the mathematician was that what Einstein wanted to do was explain the way nature worked, not merely derive beautiful equations. It was the excitement of seeing that, even by such a small effect, nature obeyed the laws he discovered in his mind that made Einstein weak with excitement.

In the same paper in which he derived the precession of Mercury's orbit, written a week before the paper that presented the final form of general relativity, Einstein made another prediction. He calculated that light would indeed bend in a gravitational field, as he had realized almost a decade earlier, but that the actual magnitude of the bending would be twice as large as he had previously estimated, and twice as large as the value one might get by simply pretending that light had mass and then using Newton's theory to calculate the effect of gravity on its trajectory. He thus predicted that light passing near the sun would be deflected by approximately 1/2000 of a degree.

As small as this value was, its predicted effect would be measurable, as Einstein realized as early as 1911, when he was still in fact predicting the wrong value. If one observed stars near the sun during a solar eclipse, their position would be shifted by this very small amount compared to where one would otherwise predict them to lie. Fortunately for Einstein, war and other human idiocies prevented a successful eclipse expedition to test his ideas until three years after he had indeed made the correct prediction. In November 1919, two British expeditions reported on their observations of a May 1919 eclipse: Einstein, not Newton, was correct.

This discovery forever changed Einstein's life and, with it, the world of physics. News of the eclipse observations spread across the headlines of papers throughout the world, and within weeks, Einstein attained a celebrity that would remain with him for the rest of his life. Special relativity had made him famous among physicists and perhaps even among educated intellectuals; general relativity made him a household name. His discovery that we are living in a possibly curved three-dimensional space had an immediate popular impact that might be akin to the revelation in Renaissance Europe that the earth wasn't flat. In a single moment, everything changed, and Einstein's fame would soon rival that of Columbus.

Part of the reason for his fame was surely the fact that he had now supplanted Newton as the father of gravity. But I think the general excitement that greeted his discovery was more deeply based, and for good reason.

While special relativity had connected space and time in a new way that made separate measurements of length and time observer dependent, space-time itself nevertheless remained a fixed background in which the events of the universe played out. In Einstein's general relativity, however, space and time become truly dynamic quantities. They are no longer mere backdrops in which the drama of life ensues, but *respond* to the presence of matter and energy, bending, contracting, or even expanding in the presence of appropriate forms of matter or energy.

One of the predictions of general relativity that took almost half a century to verify empirically was that clocks tick more slowly in a gravitational field. Normally the effect is truly minuscule, and to measure it required careful optical techniques, unstable radioactive compounds, and ultimately the use of atomic clocks.

However, sometimes, if we take into account the fact that we live in a large universe, small effects can be magnified tremendously. One of my favorite examples of this (for reasons that will become obvious in a moment) involves some work a colleague of mine and I did shortly after the discovery, on February 23, 1987, of an exploding star on the outskirts of our galaxy, the first such event seen in almost four hundred years. Its demise was observed both via the light emitted by the star, which shined with a brightness approaching that of a billion stars for days, and also via the almost simultaneous detection of ghostlike elementary particles called neutrinos, which are in fact the dominant form of radiation emitted by exploding stars. Within a few weeks of the event, there were literally scores of scientific papers (including some by me) analyzing every aspect of these signals.

About two months after this flurry, Scott Tremaine (who is now at Princeton, but at the time was at the Canadian Institute for Theoretical Astrophysics in Toronto) and I were at a meeting in Halifax, Canada, when we suddenly realized that one could calculate the extra time it would have taken for both the light and the neutrinos to travel from the distant star to Earth, due to the fact that both bursts were traveling in the gravitational field of our galaxy and hence not in a flat background. The result surprised both of us: The gravitational time delay was about six months. If it hadn't been for the warping of both space and time as predicted in general relativity, Supernova 1987a, as it became known, would have been called Supernova 1986d, as it would have been observed sometime around the middle of the previous year!

It is virtually impossible for us, who are confined to live within a curved three-dimensional space, to physically picture what such a curvature implies. We can intuitively grasp a curved two-dimensional object, such as the surface of the earth, because we can embed it in a three-dimensional background for viewing. But the possibility that a curved space can exist in any number of dimensions without being embedded in a higher-dimensional space is so foreign to our intuition that I am frequently asked, "If space is curved, what is it curving *into*?"

There are, however, mathematical ways to define the geometry of a space without the existence of extrinsic quantities. The simplest example involves something with which we are all familiar. Consider a triangle drawn on this piece of paper.

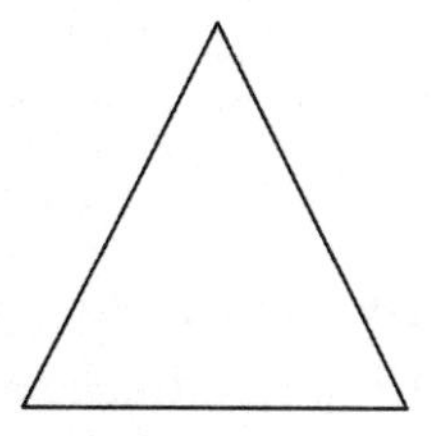

As any European high school student could tell you, the sum of the angles inside this triangle is 180 degrees, independent of the shape or size of the triangle.

Now, however, consider the following figure:

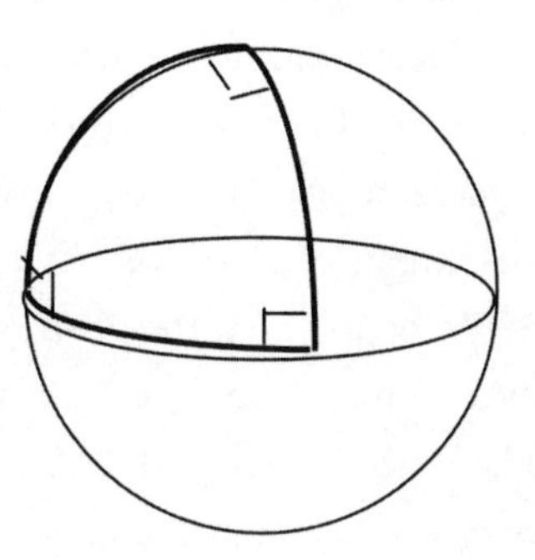

All three angles of this triangle are right angles, adding up to a sum of 270 degrees. Were we intelligent ants living on this curved surface, even if we could never circumnavigate it or view it from above, by drawing a large enough triangle and measuring the sum of its internal angles, we could nevertheless infer that we were living on a spherical surface.

Another factor distinguishes a sphere from a flat piece of paper, which

I alluded to earlier. Lines of longitude, extending from the North to the South Pole, are all parallel lines, yet all of these lines meet at both poles:

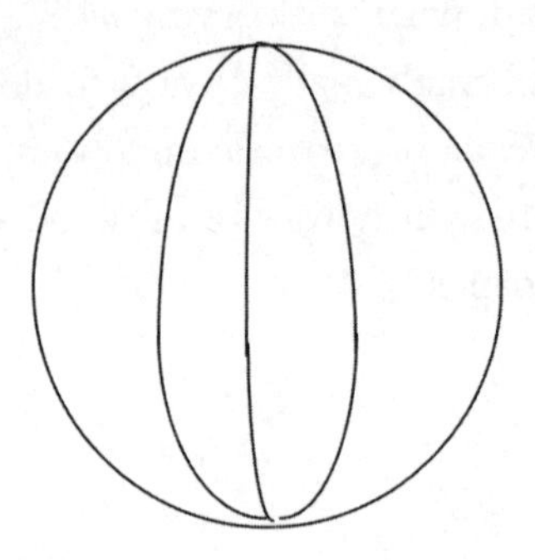

As obvious as all this might seem in retrospect, the notion that it might be possible to have a geometry where Euclid's axioms about parallel lines or about the sum of angles in a triangle might not hold caused a revolution in philosophy. Euclid's axioms had remained unchallenged for two thousand years when the mathematicians Gauss, Lobachevsky, and Bolyai independently discovered between the years 1824 and 1832 that one could build a consistent mathematical framework in which the axiom about parallel lines could be violated. So great was the resistance to these notions that the famous physicist Helmholtz felt it necessary to incorporate precisely the examples I have given here in his 1881 *Popular Lectures on Scientific Subjects* (published three years before *Flatland*), in which he described a hypothetical world of two-dimensional beings living on the surface of a sphere, in order to convince people that the abstract mathematical notions of Gauss and others could be manifested in a consistent physical reality.

Interestingly enough, both Gauss and Lobachevsky realized that if non-Euclidean geometry was possible in principle, it might also be possible in practice, and both conducted independent experiments to see if our three-dimensional space might be curved. Gauss was more modest in his attempts, merely measuring the sum of the angles in a large triangle formed by three distant mountain peaks. Lobachevsky, in contrast, performed a far more modern experiment. He observed the parallax of various distant stars, that is, the angle by which they shift compared to background objects when the earth is on one side of the sun, compared to when the earth is on the other side of the sun a half-year later. Plane geometry gives a straightforward prediction for what this shift should be for stars at a fixed distance, at least in a flat space. The shift would be different, however, if space was curved.

Given the limited sensitivity of their observations, neither Gauss nor Lobachevsky was able to obtain any evidence whatsoever for the non-Euclidean nature of space. That evidence would have to wait for almost a century, until after Einstein had made it clear what to look for.

While the British solar eclipse expeditions were able to detect the curvature of space in the vicinity of the sun, general relativity posed a much, much bigger challenge. This was a theory not merely of how objects might move throughout space and time, but of how space and time themselves might evolve. Einstein opened up the possibility of describing the dynamics of the universe itself, and since general relativity is a geometric theory, the central question of twentieth-century cosmology soon became: Is the geometry of the universe, on its largest scales, described by Euclid?

CHAPTER 6
THE MEASURE OF ALL THINGS

I believe with Schopenhauer that one of the strongest motives that leads men to art and science is escape from everyday life with its painful crudity and hopeless dreariness from the fetters of one's own everyday desires. . . . A finely tempered nature longs to escape from personal life into the world of objective perception and thought.

–Albert Einstein

I am not sure that I completely agree with Einstein's romantic view of the scientific (or artistic) enterprise, having always felt that these activities, as human endeavors, are intimately connected with the rest of our existence, dreary or otherwise. But perhaps that is one of the many reasons why Einstein was Einstein, and I am me. In any case, for over twenty years I have devoted most of my scientific effort to questions about the origin, nature, and future of our expanding universe that are about as far removed from the world of my everyday experience as can be. While I like to think that my contributions have helped us move forward in our knowledge, nothing truly prepared me for the revolutionary developments of the past decade, which is why I want to make a brief digression from the historical presentation thus far, and jump to a present-day result that has finally addressed the question first asked by Einstein almost hundred years ago.

I remember, from the time I taught at Yale, a conversation with a senior member of its astronomy department, Gus Oemler. I used to visit him regularly with crazy ideas about how one might measure such fundamental quantities as the expansion rate and the geometry of the universe. With his wealth of experience, he brought valuable skepticism to any discussions we had.

On this particular day we were discussing how to measure what has become known as the Hubble constant, a quantity that describes the expansion rate of our universe and which, in a manner characteristic of much of astronomical nomenclature, is actually not in general a constant quantity at all, but varies over cosmological time. In any case, in the course of our talk Gus revealed to me a theorem he had about the universe: "I believe that the universe will always conspire to make any fundamental and precise measurement of cosmological quantities such as the Hubble constant impossible."

As outrageous as this claim might seem, it was based on decades of experience in astronomy. On many occasions over the past thirty to forty years, astronomers had claimed to make definitive measurements about such quantities as the curvature of the universe or its expansion rate, and each time it turned out that subtle uncertainties that had not been anticipated by the observers clouded their results, ultimately invalidating many of them.

Thus it was that in 1999 I was unprepared for a totally clean and unambiguous observation, using a method that I had in fact written about in a slightly different context almost a decade earlier: a profound and direct measurement of the geometry of the universe in which we live. Equally remarkable is the fact that the method used is almost identical, at least in principle, to that used by Lobachevsky over 150 years earlier to first explore for a possible curvature of space. The only difference is that the triangles we now use "as reference points" span not the distances to the nearest stars, but rather the distance across the entire visible universe.

This observation because possible because of the accidental discovery, forty years ago, of a then-mysterious thermal bath of radiation bombarding us from all directions, with a temperature of about three degrees above absolute zero (on the Kelvin temperature scale, in which absolute zero, the coldest temperature possible, is labeled zero, unlike the Fahrenheit scale, where absolute zero is about minus 450 degrees). It didn't remain mysterious for long, however. When the perplexed scientists at Bell Laboratories who had found this excess "noise" in their antennas went down the road with their findings to Princeton University, the scientists there informed them that they had discovered the afterglow of the big bang.

Shortly after Edwin Hubble's discovery in 1929 that the universe is expanding, it was realized that by following this expansion backward in time one might hope to trace out the thermal history of the universe. By going back over ten billion years, the universe one would encounter would have consisted of a hot, dense gas of particles and radiation in thermal equilibrium. Such an extrapolation was, of course, bold, but it did make many theoretical predications possible, all of which could be tested against observations. The most robust of them, perhaps, involved the prediction of a background of radiation left over from the big bang that would have permeated the universe, and would have been cooling as the universe expanded over the billions of years between the big bang and now.

We can understand why this microwave radiation bath exists and what its origin is by remembering one of the fundamental facts of electromagnetism: Light travels at a finite velocity through space, so that the farther out we look in the universe, the further back in time we are looking. Every time we peer through a telescope, we are doing cosmic archaeology.

Pushing this idea to its logical limit means that in principle, if the universe had a beginning a finite time ago in the past, if we look out far enough with sufficiently powerful telescopes, we should be able to witness the big bang itself! Unfortunately, however, there is a fundamental roadblock to actually achieving this goal. Between the big bang and now, the universe went through an opaque period when it was so hot and dense that light could not travel unimpeded throughout space, unlike the present time, when it can traverse the vast distances between stars and galaxies.

Using well-known laws of physics, we can actually calculate the precise time before which the universe was opaque. At a temperature of greater than about three thousand degrees Kelvin above absolute zero the ambient radiation present is sufficiently energetic to break apart the bonds that hold atoms such as hydrogen together. Hydrogen is the simplest atom, made up of a single proton, surrounded by an electron. At extremely high temperatures, absorption of energy from a radiation bath is sufficiently great to allow the electron to be knocked free of its electronic bond to its host proton. While it could be captured again by another bare proton, the radiation would once again knock it free. At the high temperatures of the early universe, therefore, hydrogen was ionized, meaning that its charged particles (protons and electrons) were separated and not bound together into neutral atoms.

Now, ionized matter, being charged, interacts very strongly with electromagnetic radiation. Thus, a light ray cannot permeate a configuration of ionized atoms, which we call a plasma, without being constantly absorbed and reemitted. This means that as we attempt to look back farther and farther we eventually hit a metaphorical wall. If we try to look back to earlier times, we simply cannot do so using electromagnetic radiation, just as we cannot look behind the walls in the room that surround us, because the radiation cannot penetrate their surface. Indeed, when we look at a wall, we are seeing radiation that has been absorbed at the surface, and later reemitted into the room, making its way through the transparent air to our eyes.

Similarly, as the universe cooled below three thousand degrees, and neutral atoms could finally form, space became transparent to radiation. Thus, we should expect to be able to see a "surface" located billions of light years away from us that represents the time when the universe first became neutral, when it was about three hundred thousands years old. From this surface we should expect to receive a bath of radiation coming at us from all directions. Since the universe has been expanding and cooling since the time that that surface originally emitted the radiation, by the time it gets to our sensors the radiation should have cooled considerably.

The first people to propose that such a radiation background should exist were a research group associated with the scientist and writer

George Gamow, whose many popular books inspired generations of young people (including me) to think about science. At the time that Gamow's colleagues Robert Alpher and Robert Hermann made their proposal, no one really took the big bang picture seriously. However, as I mentioned previously, twenty years after his prediction two young would-be radio astronomers at Bell Laboratories in New Jersey discovered an unusual source of noise in a sensitive radio receiver they planned to use to scan the heavens. The noise was characteristic of a background of radiation at a temperature of about three degrees above absolute zero. While they had no idea of it at the time, this was more or less precisely the temperature such a radiation bath remnant of the big bang was predicted to now possess.

Because this radiation emanates from within the first three hundred thousands years after the big bang, it has become one of the most important probes of cosmology. By carefully measuring its properties, one can hope to glean a wealth of information about the early universe.

In 1999 an experiment was launched near the South Pole to measure this background radiation with unprecedented accuracy. A microwave radiation detector was set aloft on a huge balloon that would rise to a hundred thousand feet above the earth, well above most of the atmosphere that would otherwise absorb some of the radiation before it could reach the earth. The balloon with its important payload took almost two weeks to circle Antarctica, returning close to the spot from where it had been launched (which is why it was called the boomerang experiment), and during this time the microwave radiometer focused on a small patch of the sky, measuring the temperature of the background radiation across the patch to an accuracy of better than one part in one hundred thousand.

What the experimenters who built and operated the device were looking for was a very particular distribution of hot and cold spots about one degree across in the microwave sky. This angular size has a special significance, for it represents the distance light could have traveled across points on the "surface" from which the microwave background emanates, about three hundred thousand years following the big bang.

Since no signal can be transmitted faster than light, this distance, about three hundred thousand light years, thus represents the largest distance over which the effects of any physical disturbance located at one place could propagate.

Put another way, this scale is the largest scale over which local physical processes could respond to macroscopic conditions. For example, a bit of excess mass in some region might, by its gravitational self-attraction, begin to collapse. The increased density in this region would then cause a corresponding increase in pressure. Such effects of pressure responding to gravity could only occur across regions smaller than or equal to three hundred thousand light years across, however, because on larger scales lumps of excess mass do not even know they are lumps–light cannot have traveled across them. This is why the angular scale associated with this distance is special–it is associated with the largest size regions within which there is causal contact. For this reason, one would expect to see a residual imprint on the microwave background on such scales.

Such a situation in principle provides us with all the ingredients we need to be able to directly probe the geometry of the universe, by giving us a large triangle, as shown below. Two of the sides of the triangle represent the distance from Earth out to the surface from which the microwave background emanates. The third side is this special distance across the surface, representing the maximum distance a physical signal could have propagated at that time, about three hundred thousand light-years.

General relativity implies that light rays travel in space in straight lines, but if the underlying space is curved, the trajectories of the light rays themselves will be curved. Thus, the light rays emanating from the edges of a region spanning a distance of three hundred thousands light-years across would follow one of three different kinds of trajectories on their way to the earth. If the universe is positively curved, then the light rays would bend inward on their travels. If it were negatively curved, the light rays would bend outward. And if the universe is flat, the light rays would follow straight lines.

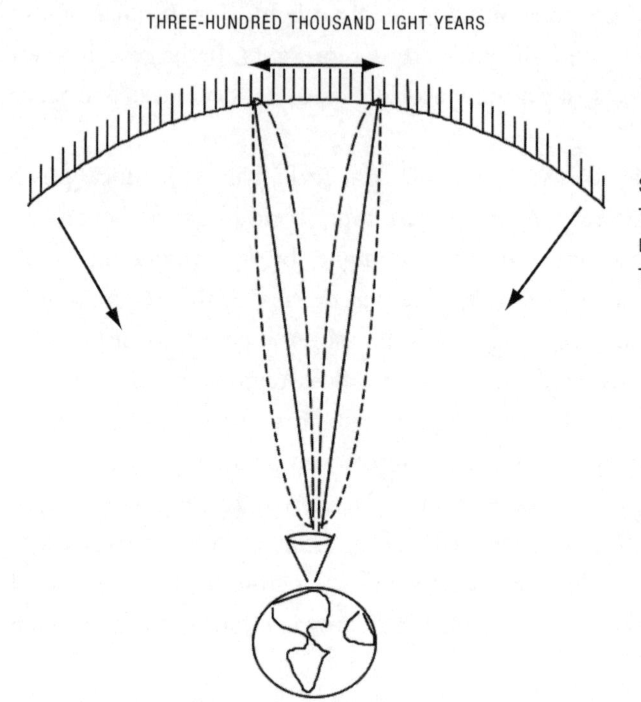

From the point of view of an observer on Earth, then, the angular size of these regions will depend upon what the geometry of the universe is. If space is negatively curved across the universe, the apparent angular size of these hot spots and cold spots will be reduced. If it is positively curved, the hot and cold spots will appear enlarged. If it is flat, the size will be somewhere in between.

In 1999 the boomerang experiment released its results, with complex charts demonstrating the quantitative features of the temperature variations across the region of the microwave sky that it observed. However, in the spirit of the statement that a picture is worth a thousand words, the experimental team also produced a graphical representation of their findings. Here is an actual false color image of the data, with hot spots one shade and cold spots another, compared to three computer-generated versions of what you might expect for a positively curved, flat, and negatively curved space.

Here, for the first time in human history, was an empirical observation capable of disentangling the geometry of the entire visible universe. And

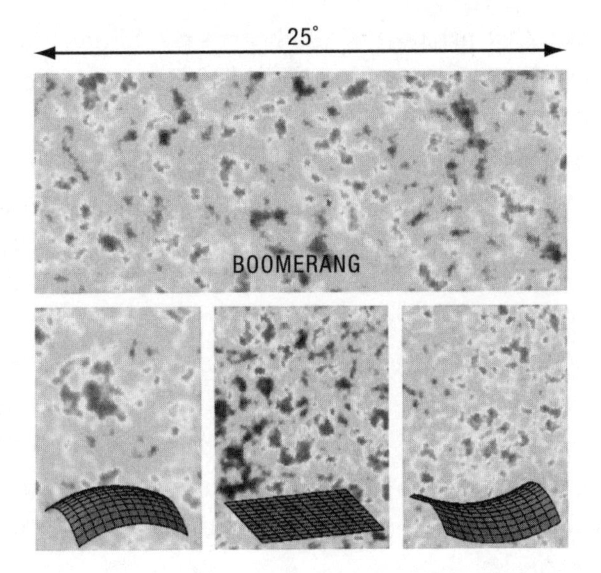

you don't have to be a rocket scientist to discern the answer. As in the Goldilocks story, the lumps predicted in the positively curved universe were too large compared to the observations, while the lumps in a negatively curved universe were be too small. A precisely flat universe, however, would produce more or less precisely what was observed.

Just as Lobachevsky had inferred 150 years earlier, on a scale that we now recognize would have been far to small to detect the minute curvature of space that might have existed on these scales, observations of the cosmic microwave background have now convincingly suggested that we live in a flat universe.

One's first response might be "How boring." Of all the interesting possible universes to live in that are allowed by general relativity, why should we live in one that is precisely flat on large scales?

Before I attempt to answer that, let me attempt to clear up a possible misconception that you may have arrived at from what you have just read. Remember that I described earlier how Einstein's theory of general relativity was first experimentally confirmed in 1919 by witnessing the fact that light rays bent around the sun. Yet I have now just argued that light rays that traverse the universe travel in straight lines.

These two facts are not inconsistent. Matter can locally curve space in its vicinity, as the sun, the earth, and even you do. However, the fundamental question that has puzzled physicists since Einstein first proposed his theory was whether the sum total of all the matter and energy in the universe produces a net curvature of space on the largest scales. If it did, one could imagine, for example, as Einstein first did, that space could ultimately curve back upon itself so that one could live in a finite universe, but one without end. Namely, if you looked far enough in any direction, you would see the back of your head! It is like a three-dimensional version of living on the surface of an expanding balloon.

A finite but endless universe is fascinating, but it does have one drawback. If matter and radiation are all that make up such a universe, general relativity implies that it must ultimately recollapse back into a hot, dense reverse of the big bang. This provides a rather unpleasant end, and so it is fortunate that other possible geometries for the universe exist that may imply less violent finales.

A negatively curved universe, like a three-dimensional version of a horse's saddle, can be infinite in spatial extent, and such a universe containing matter and radiation will expand indefinitely. With time the universe would cool down, its stars would ultimately burn out, and it would become cold and dark. This, too, is not a particularly pleasant future, but the timeframe over which the darkness would fall is so gradual–trillions of years–that such a universe, which ends with a whimper rather than a bang, seems more hospitable, at least from a human perspective.

Falling right between these two extremes is a flat universe. In such a universe containing matter and radiation, our expansion will continue to slow with time, but it will never quite stop. Like a negatively curved universe, it, too, can be infinite in spatial extent. However, because the expansion rate slows more quickly in this universe than in a negatively curved space, the time it takes before such a universe becomes cold and empty is far longer.

Longevity is not the reason that theorists preferred a flat universe long before observations confirmed this to be the case, however. The reason for their preference is partly aesthetic and partly practical. Einstein's equations from general relativity establish a relationship between the curvature of the universe, the rate of its expansion, and the total density of matter and energy within it. Observations of the expansion of the universe and measurements of the total matter density had long established that these quantities were within an order of magnitude or so of what was required to produce a flat universe.

Now, being within an order of magnitude is certainly not compelling evidence, on its own, of equality. But a remarkable mathematical relationship does exist that made believers out of many theorists long before the appropriate experimental evidence was amassed. It turns out that general relativity implies that if the geometry of the universe is not flat, then, as the universe expands, it quickly moves farther and farther away from the mathematical equality implied by flatness. Since the universe is over ten billion years old, it is difficult to imagine how the relation between the expansion rate and the mass density could still remain so close to that for a flat universe unless the universe was, in fact, precisely flat.

This puzzle was so significant that cosmologists even gave it a name: the flatness problem. In 1981 a Stanford physicist (now at MIT) named Alan Guth proposed an ingenious mechanism that would resolve this puzzle by producing, independent of its initial conditions, a flat universe today. His idea, called inflation, was that the universe underwent a rapid early period of expansion, far faster than had previously been envisaged. Like a balloon being blown up, as the universe inflated, any original curvature of space would be progressively reduced, ultimately producing a universe that was indistinguishable from a precisely flat universe.

What's more, Guth demonstrated that physical conditions that would lead to an early inflationary phase could arise naturally in so-called grand unified models of particle physics, which I shall later describe, in which the fundamental forces in nature are unified into a single force at very early times. Once Guth had shown that inflation could easily result in these models, and how it could resolve a variety of fundamental problems in cosmology beyond the flatness problem, it quickly became the basis of what is now considered the standard model of cosmology.

Aside from Guth's inflationary paradigm, there is, however, another reason a flat universe is particularly attractive, at least from a theorist's perspective: The total gravitational energy of a flat universe is precisely zero!

How can a universe full of matter and radiation have zero total energy? While, the energy associated with these quantities, in the absence of considerations of gravity, is indeed positive, it turns out that the gravitational energy of attraction between objects is negative. This is another way of saying that it takes energy to pull objects farther apart, so they have less energy if they are close together. Hence, all objects of a finite size have less gravitational energy than they would have if they were dispersed over infinitely large distances. If we define such a state in which matter is infinitely diluted as having zero energy, then all other, smaller, configurations have negative energy. If this negative energy precisely cancels the positive energy of matter and radiation in the universe, then general relativity tells that the overall curvature of space vanishes.

Moreover, with zero net energy, the possibility that the universe itself arose spontaneously out of nothing becomes at least plausible, since one would imagine that "nothing" would also have zero energy. As Guth put it: "There is such a thing as a free lunch!"

It was theoretical considerations such as these, which are primarily mathematically aesthetic, that convinced most theorists and ultimately even many observers, well in advance of the cosmic microwave background observations, that the universe was flat. In this case, as sometimes but not always happens in science, nature cooperated.

However, it was premature to slap ourselves on our collective backs and congratulate one another. For, what actually makes the universe flat is something that no one, or at least almost no one, anticipated. Perhaps the most puzzling discovery in all of physics during the past century has been the fact that the dominant form of energy in the universe is not associated with matter or radiation at all. Rather, it appears that empty space, devoid of any particles at all, carries energy–enough energy, in fact, to overwhelm, by a factor of almost three, the energy of everything else in the universe.

This energy of empty space, sometimes called "vacuum energy" or "dark energy," is the most mysterious form of energy we know of. No one currently has a good explanation of why empty space should have precisely this amount of energy, and, as we shall see, trying to understand its nature is currently driving much of our current scientific thinking about the nature of space and time itself.

The discovery of a mysterious energy permeating all of empty space also changed everything in the way we think about cosmology. Even the original, vital connection between geometry and destiny is now gone. If empty space can possess energy, a positively curved universe need not ultimately collapse, while a negatively curved or flat universe need not expand forever.

Still, as I have suggested, it could be that there might be some deeper connection between the geometry of space and its energy content, perhaps something that involves probing yet deeper into the meaning of space and time. Certainly the puzzle of dark energy is so revolutionary it motivates even extreme reconsiderations of that nature of space and time. The resolution of this mystery may not be as revolutionary as the question itself, but one never knows until one explores the possibilities.

But extraordinary claims require extraordinary evidence, as Carl Sagan used to say. We shall return to this mystery later in the book. First, however, we shall explore how the collective creative imagination of the world responded to our first revolution in the physics of space and time inspired by Einstein and later Minkowski: Namely, the existence of a four-dimensional space-time continuum associated with special relativity.

CHAPTER 7

FROM *FLATLAND* TO PICASSO

Ever drifting down the stream–
Lingering in the golden gleam–
Life, what is it but a dream?

–Lewis Carroll, *Through the Looking Glass*

While life may imitate art, it is nevertheless also true that art imitates life. One might thus wonder whether the publication of Abbott's *Flatland* within a decade following Maxwell's discovery about the nature of otherwise invisible electric and magnetic fields and less than a decade before Michelson and Morley's experiments to probe the ether and Lorentz's pioneering speculations about the nature of space and time was purely a coincidence. Was there something in the intellectual air at the time that suggested something revolutionary was about to occur in our understanding of nature?

In one sense the answer to this question is clearly no. It was, after all, in 1900 that Lord Kelvin uttered his famous remark that all laws of physics had already been discovered and all that remained were more and more precise measurements. Yet in spite of such hubris, scientific and mathematical puzzlement about the nature of space and time had been spilling over to the literary imagination for well over a century before Abbott wrote his story.

The notion that time might somehow be considered a fourth dimension actually appeared in print as early as 1754, in an article by Jean Le Rond d'Alembert on "Dimensions" in his *Encyclopédie,* although he attributed the idea to a friend, possibly the French mathematician Joseph-Louise Lagrange. A hundred years later German psychologist and spiritualist Gustav Fechner wrote a satirical piece involving a "shadow man," the shadow projection of a three-dimensional image. Interestingly, Fechner argued that such shadow figures would interpret the effects of motion perpendicular to their plane of existence (which they, of course, could not perceive as movement in space) as acting like time. Fechner's combined interest in extra dimensions and spiritualism presaged, as we shall see, events that would unfold a half a century later.

Ultimately the notion of time as a fourth dimension was made famous within popular culture a full decade before Einstein's special relativity and thirteen years before Minkowski clarified the dimensional relationship between space and time by none other than H. G. Wells in his classic science fiction epic, *The Time Machine,* published in 1895. On the very first page of this novel, Wells's hero, the Time Traveller, has the following dialogue with an audience he has invited for the occasion:

> "You must follow me carefully. I shall have to controvert one or two ideas that are almost universally accepted. The geometry, for instance, they taught you at school is founded on a misconception."
>
> "Is not that rather a large thing to expect us to begin upon?" said Filby, an argumentative person with red hair.
>
> "I do not mean to ask you to accept anything without reasonable ground for it. You will soon admit as much as I need from you. You know of course that a mathematical line, a line of thickness NIL, has no real existence. They taught you that? Neither has a mathematical plane. These things are mere abstractions."
>
> "That is all right," said the Psychologist.
>
> "Nor, having only length, breadth, and thickness, can a cube have a real existence."

"There I object," said Filby. "Of course a solid body may exist. All real things."

"So most people think. But wait a moment. Can an INSTANTANEOUS cube exist?"

"Don't follow you," said Filby.

"Can a cube that does not last for any time at all, have a real existence?"

Filby became pensive. "Clearly," the Time Traveller proceeded, "any real body must have extension in FOUR directions: it must have Length, Breadth, Thickness, and–Duration. But through a natural infirmity of the flesh, which I will explain to you in a moment, we incline to overlook this fact. There are really four dimensions, three which we call the three planes of Space, and a fourth, Time. There is, however, a tendency to draw an unreal distinction between the former three dimensions and the latter, because it happens that our consciousness moves intermittently in one direction along the latter from the beginning to the end of our lives."

"That," said a very young man, making spasmodic efforts to relight his cigar over the lamp, "that . . . very clear indeed."

"Now, it is very remarkable that this is so extensively overlooked," continued the Time Traveller, with a slight accession of cheerfulness. "Really this is what is meant by the Fourth Dimension, though some people who talk about the Fourth Dimension do not know they mean it. It is only another way of looking at Time. THERE IS NO DIFFERENCE BETWEEN TIME AND ANY OF THE THREE DIMENSIONS OF SPACE EXCEPT THAT OUR CONSCIOUSNESS MOVES ALONG IT. But some foolish people have got hold of the wrong side of that idea. You have all heard what they have to say about this Fourth Dimension?"

"I have not," said the Provincial Mayor.

"It is simply this. That Space, as our mathematicians have it, is spoken of as having three dimensions, which one may call Length, Breadth, and Thickness, and is always definable by ref-

> erence to three planes, each at right angles to the others. But some philosophical people have been asking why THREE dimensions particularly–why not another direction at right angles to the other three?–and have even tried to construct a Four-Dimension geometry. Professor Simon Newcomb was expounding this to the New York Mathematical Society only a month or so ago. You know how on a flat surface, which has only two dimensions, we can represent a figure of a three-dimensional solid, and similarly they think that by models of three dimensions they could represent one of four–if they could master the perspective of the thing. See?"

This passage is remarkable not merely because of Wells's anticipation of a connection between space and time in a four-dimensional framework, but because he correctly recognized that what fascinated writers and the public alike was not a *temporal* fourth dimension but a spatial one.

Wells also wrote several stories reminiscent of *Flatland,* in which he utilized four spatial dimensions as plot devices. In no fewer than four tales Wells exploited different manifestations of extra dimensions that would be borrowed by a host of future science fiction writers. These included a story involving a person being turned into his mirror image through a four-dimensional rotation, the possibility of connecting otherwise distant locations in three-dimensional space via a four-dimensional portal, the mysterious appearance and disappearance of a four-dimensional being (an angel, as it happens) traveling through our three-dimensional plane of existence, and finally an object achieving invisibility by sliding into the fourth dimension.

About 150 years earlier, around the same time as d'Alembert was writing, none other than Immanuel Kant was pondering the possibilities of extra spatial dimensions. While he may have felt that Euclidean geometry was an essential part of existence, he was much more sanguine about variations beyond our three-dimensional space, although he felt that while they could exist, they must be separate from ourselves. He discussed this possibility in his very first published work, *Thoughts on the True Estimation of Living Forces,* concluding: "Spaces of this kind, however, can not stand in connection with those of a quite different constitution. Accordingly such spaces would not belong to our world, but must form separate worlds."

The German physicist and mathematician August Möbius, father of the famous one-sided Möbius strip, followed up on Kant's earlier musings from the 1700s and came up with an interesting suggestion. He argued in 1827 that a fourth dimension would allow otherwise distinct three-dimensional figures–such as a right hand and a left hand–to coincide. Namely, just as a mirror flips left and right, one could turn a right hand into a left hand by twisting it into a fourth dimension and back again. Indeed, Kant, himself, in his *Prolegomena to Any Future Metaphysics* (1793) wondered explicitly about how a right hand becomes a left hand when viewed in a mirror, and so two identical objects can at the same time be completely different.

The premise inherent in *Flatland* was that we could simply be ignorant of an ever-present fourth spatial dimension, which would appear as foreign to our intuition as a third dimension would be to a two-dimensional being. Abbott was, of course, not writing in a vacuum, and there was a swirl of activity in England in the years prior to 1884 surrounding attempts to understand physically and mathematically what a fourth dimension might be like.

As I have mentioned, H. G. Wells himself wrote at least one tale in which this very issue is central. His "The Plattner Story"(1896) focuses on an individual who moves into the fourth dimension and returns with left and right inverted. Almost eighty years later, a charming rendition of this same apparent paradox was replayed in Lars Gustafsson's tale *The Death of a Beekeeper.* The protagonist muses: "But, since I moved outside the normal dimensions, right and left somehow got exchanged. My right hand is now my left one, my left hand my right one." At the same time, this transition changes his previous, pessimistic, view of our world: "Returned into the same world and see it now as a happy one. The shreds of peeled paint on the door belong to a mysterious work of art." If only it were so.

Numerous authors before Abbott had exploited two-dimensional beings as an allegory to help us imagine a fourth dimension. In England, the mathematician J. J. Sylvester wrote a popular article using them in 1869. In it he quoted from the biography of the great mathematician Gauss, in which the late mathematician was reported to have stated that he had kept several geometric questions aside, waiting to pass on so that he would have a better appreciation of four or more dimensions!

Not only was Sylvester a bold advocate of understanding four dimensions, he also firmly believed that higher dimensions actually exist, and strongly asserted an "inner assurance of the reality of transcendental space."

Another mathematician who popularized two-dimensional beings was Charles Dodgson, known to the world as Lewis Carroll, the author of the Alice in Wonderland stories. In an 1865 story, entitled "Dynamics of a Parti-cle," he described a romance between a pair of linear, one-eyed animals moving along on a flat surface.

When I first learned this fact I was particularly intrigued, because *Through the Looking Glass* (1872) was the first story I could remember that envisaged a foreign world lying right beneath our eyes. Moreover it was a world I had been fascinated with as a child–so much so that it influenced the title of this book. What if the world hiding on the other side of a mirror was real?

I have since learned, however, that Dodgson was in fact parodying the British fascination of the time with the literal idea of a fourth dimension. Dodgson's mirror world of talking chessmen and tiger lilies may not appear to a modern reader to deal directly with such issues, but apparently the psyches of nineteenth-century British readers were more attuned to his satire. At least the white queen's memories involved both the past and the future, so time appeared to be heavily involved in the mix. Or maybe it was the queen's propensity for believing six impossible things before breakfast that Dodgson employed to parody the fads of the time. Actually, Dodgson later became interested in the occult, and with that presumably his skeptical attitude toward extra dimensions disappeared.

In the late 1870s a more sinister application of the fourth dimension appeared when a German physicist and astronomer, J. C. F. Zöllner, who in his day job (or, more appropriately, night job) actually invented a method of accurately measuring the brightness of stars, became fascinated with an American medium named Henry Slade. In séances carried out for Zöllner and others Slade performed magic tricks–such as untying a knotted cord without touching it, and transporting objects out of a sealed container—that seemed to defy explanation unless somehow he was reaching "into" an extra dimension. Like many of those who followed (including Russell Targ and Harold Putoff, who in the 1970s claimed to have found scientific evidence for remote perception), Zöllner left scientific skepticism behind and fell for Slade's chicanery, becoming his ardent defender and writing prolifically about Slade's empirical demonstrations of the existence of extra dimensions.

Zöllner's fascination with Slade and the occult strikes to the heart, just as forcefully as Alice's yearning to disappear into the mirror, of why humans have always seemed to want to believe in the possibility of extra dimensions. We seem to need somewhere beyond the world of our experience, a place that's either better or just different.

Part of this desire, I believe, arises because, while science describes the workings of the natural world, it does so without reference to "purpose," so that even those who adhere to scientific principles may ultimately find its view of reality lacking. For Zöllner and others, the fact that the possibility of the existence of extra dimensions was at least allowed by science—even if no direct evidence had been forthcoming—also meant it allowed a place for a world of purpose, the spiritual world, to exist.

This deep yearning is undoubtedly associated with ubiquity of religion in the human experience, of which I spoke at the beginning of this book. The need for a hidden god to guide the universe of our experience while existing outside of that universe, and the hoped-for existence of a "better place" where we might go after we die, are part and parcel of the same sense of longing for something transcendent that is evident even in the fourteen-thousand-year-old cave paintings in France.

Actually, an explicit connection between the spiritual realm and something akin to extra dimensions predates Zöllner by at least two centuries. In 1671 the Cambridge philosopher Henry More proposed that spirits were four-dimensional. He even framed a pseudoscientific quantity, which he called "spissitude," which differentiated between identical bodies of living and dead persons. Living ones had more spissitude, which was nevertheless unobservable because it had thickness in the fourth dimension.

A variation on this theme was taken up a century later by the Swedish scientist, linguist, theologian, and mystic Emanuel von Swedenborg. Swedenberg wrote over fifty works on science, chemistry, and theology and was fluent in eleven languages. His remarkable combination of brilliance, spiritual flights of fancy, and mystical visions (which may have been a symptom of underlying schizophrenia) had a huge impact on generations of writers from Goethe to Henry James. In midlife, inspired by a series of visionary experiences, he abandoned his scientific investigations and devoted himself to prophecy and spirituality. Perhaps because of his scientific background, however, we find in Swedenborg an explicit turning to science as an explanation of the spirit world. In particular, he argued that humankind existed simultaneously in two parallel worlds, the material and the spiritual, the latter of which is populated by angels and also humans after they die. Among his other postvisionary writings is the book *Earths in the Universe,* in which he claimed that the moon was populated by aliens who speak through their stomachs with a language that sounds like belching!

By the time of *Flatland* many English clergy had taken up with renewed fervor the notion that the fourth dimension was associated with spritual phenomena. Their viewpoint was presented by A. T. Schofield in his book, *Another World* (1888), in which he wrote: "We conclude, therefore, that a higher world than ours is not only conceivably possible, but probable: secondly, that such a world may be considered a world of four dimensions, and thirdly, that the spiritual world agrees largely in its mysterious laws . . . in its miraculous appearances . . . with what would be the laws, language, and claims of a fourth dimension."

An interesting and similar view is expressed in a piece by N. A. Morosoff, "Letter to my Fellow-Prisoners in the Fortress of Schlusselburg" (1891), where he muses about how he and his three-dimensional friends might appear if they managed to escape and visit a nearby lake to two-dimensional beings who were confined to the surface of the lake: "In their eyes you would be an all-powerful being–an inhabitant of a higher world, similar to those supernatural beings about whom theologians and metaphysicians tell us."

A fourth spatial dimension was not just exotic but offered many possibilities that obviated the constraints of our existence, and in so doing promised to free our minds from the vicissitudes of our own tedious three-dimensional lives. What if traveling into another dimension allowed one to touch back down into our three dimensions of space, but at a different time? Would time travel then be possible? What about ESP or remote sensing? Could one somehow "sense" phenomena through perceptions into another dimension that one could not perceive otherwise? What about God, the spirit world, or even aliens? How many angels in the fourth dimension could dance on the head of a pin? As we shall later see, all of these issues have been the fodder for fiction, speculation, and belief in the twentieth century.

Abbott himself clearly viewed the fourth dimension as providing possibilities for performing precisely the kind of magic of which Zöllner believed Slade was capable. Witness his 2D hero's dialogue with his 3D spherical guide, who visited him coincidentally at midnight on the last day of 1999, which even in *Flatland* they incorrectly referred to as the end of the second millenium:

> "Pardon me. O Thou Whom I must no longer address as the Prefection of all Beauty; but let me beg thee to vouchsafe thy servant a sight of thine interior."
>
> "My what?"
>
> "Thine interior: thy stomach, thy intensines."
>
> "Whence this ill-timed impertinent request? And what mean by saying that I am no longer the Perfection of all Beauty?"
>
> "My Lord, your own wisdom has taught me to aspire to One even more great, more beautiful, and more closely approximate to Perfection than yourself. As you yourself, superior to all *Flatland* forms, combine many Circles in One, so doubtless there is one above you who combines many Spheres in One Supreme Existence, surpassing even the Solids of Spaceland. And even as we, who are now in Space, look down on *Flatland* and see the insides of all things, so of a certainty there is yet above us some higher, purer region, wither thou dost surely purpose to lead me. . . . Some yet more spacious Space, some more dimensionable Dimensionality, from the vantage-ground of which we shall look down together upon the revealed insides of Solids things, and where

> thine own intestines, and those of thy kindred Spheres, will lie exposed to the view of the poor wandering exile from *Flatland,* to whom so much has already been vouchsafed. . . . What therefore more easy than now to take his servant on a second journey into the blessed region of the Fourth Dimension, where I shall look down with him once more upon this land of Three Dimensions, and see the inside of every three-dimensioned house, the secrets of the solid earth, the treasures of the mines of Spaceland, and the intestines of every solid living creature, even of the noble and adorable Spheres. . . . I ask therefore, is it, or is it not, the fact, that ere now your countrymen also have witnessed the descent of Beings of a higher order than their own, entering closed rooms, even as your Lordship entered mine, without the opening of doors or windows, and appearing and vanishing at will?"

Magic tricks aside, Abbott's tongue-in-cheek handling of A. Square's path to enlightenment through successively higher dimensions is typical of another, perhaps more profound, aspect of the literary tradition associated with the explorations of other dimensions. This is its use in fiction as a medium of social criticism. As we have seen, Carroll may have used Alice's experiences in the looking glass house to poke fun at British idiosyncrasies, but Abbott's story is rife with implicit satire regarding racism, sexism, and even some aspects of religion. In Flatland, women are Lines, essentially the lowest form of being, who, because of the fact that they might accidentally pierce unsuspecting males, must make a special cry in all public places to make people aware of their location, and they have segregated entrances in all buildings. Triangles are the next lowest class, with very limited rights. Among them, Triangles with unequal sides are workmen, who live lives of servitude. If, by chance or careful arrangement, an Isosceles Triangle gives birth to a more prestigious Equilateral Triangle, the child is removed and sent to Equilateral parents, and forbidden from every seeing its original parents again. Squares are a bit higher in status, and so on, all the way up to Spheres, who are priests and the most exalted of all. It is heresy in Flatland to speak of higher dimensions, and one can be jailed for life for thinking or suggesting a better possible existence in three dimensions.

In *The Time Machine* H. G. Wells employed temporal travel as a means of using the future as a mirror for the present. The destruction of society by misuse of technology, which gave rise to a caste system populated by widely divergent biological descendants of present-day humans, allowed Wells, in a manner that has been a characteristic of much of science fiction, to explore issues that would have been more contentious if framed purely in the here and now.

Yet another aspect of the use of higher dimensions in nineteenth- and twentieth-century literature was as an impetus to free up the mind for speculation about the universe. Charles Hinton, a British mathematician and physicist who essentially devoted his life to writing about the fourth dimension, took this approach. He felt that if we could improve our intuition to comprehend the nature of four-dimensional objects, our minds would be liberated to better appreciate all aspects of the world around us. To this end he wrote innumerable stories and books outlining precise methods by which he felt one could visualize objects such as the four-dimensional extension of a cube, which he called a tesseract, by picturing different three-dimensional cubes that would provide its faces, just as two-dimensional squares provide the faces of a three-dimensional cube, or by imagining how the image of such an object might be projected onto three dimensions, just as one might project the image of a three-dimensional cube onto the surface of a page.

I recommend trying the latter if you truly want to get an appreciation for how hard it must have been for A. Square to attempt to visualize a sphere. Incidentally, it was none other than Abbott's Square who presented a simple mathematical algorithm to help out. How many end points in a line? Two. How many end points in a square? Four. How many end points in a cube? Eight. It does not take a rocket scientist to extend the sequence 2,4,8, to imagine that a tesseract must have sixteen end points. Similarly, if a line is obtained by joining two points, and a square can be obtained by joining together four lines, and a cube six squares, one should be able to construct a tesseract by appropriately connecting eight cubes.

Because Hinton believed the world was, in actuality, four-dimensional–as he put it, "We must really be four-dimensional creatures or we could not think about four dimensions"–he considered how a true four-dimensional understanding might alter our scientific worldview. In this respect Hinton appears strikingly modern, and some of his ideas bear at least a resemblance to current proposals I shall later describe, even if Hinton himself had no appropriate underlying theoretical basis for his contentions. Among his proposals were the suggestion that the existence of higher dimensions might help in our understanding of minute elementary particles, whose physical extension might be as large in a fourth dimension as it is in the other three. He also wondered whether the existence of positive and negative electric charges might somehow be a reflection of some underlying four-dimensional phenomena. Finally, he considered whether the very space in which we live, which was then thought to be permeated by an invisible ether (which, as we have seen, Albert A. Michelson, a student at the Naval Academy, where Hinton taught for awhile, would soon demonstrate did not exist), might have been formed as the common boundary of two adjacent four-dimensional spaces, just as a line can form the common surface of two adjacent squares, and a plane the common surface of two adjacent cubes.

Actually a connection between the long-sought ether and a fourth dimension had another, somewhat weirder manifestation. In the 1860s William Thomson, better known as the famous physicist Lord Kelvin, proposed the interesting idea that matter is made up at a fundamental level of three-dimensional "vortex rings" in the ether. Vortex rings are like smoke rings that swirl around and around on themselves, decoupled from the air around them.

Thomson's notion was actually reasonably well founded, and, as we shall again see, seems strikingly modern. It explained, for example, why atoms would have a finite size but would nonetheless be indivisible. (If you cut a smoke ring in half, it just dissipates in the air.)

While Thomson's proposal eventually died as atoms became better understood, it did spawn a far wilder concept, which appeared in a book entitled *The Unseen Universe* (1875) by B. Steward and P. G. Tait. The latter was a very highly regarded mathematician and also a former collaborator of Kelvin. These authors, returning once again to a connection between spirits and extra dimensions, suggested that our very souls existed as knotted vortex rings in the ether. These knots, created by God, could of course only be unknotted by moving into a higher dimension. (Thomson's notions of vortices and the ether also inspired the French writer Alfred Jarry, who was connected with the cubist artists as I shall soon discuss, to write a "Commentary" on four dimensions and possible time machines.)

Steward and Tait's idea might not be worth mentioning, except for the fact that none other than James Clerk Maxwell wrote about it in his famous 1876 *Britannica* encyclopedia article on the ether. He was apparently so amused by it that he also wrote a poem about unknotting his soul in four dimensions, which you can find quoted later in this book.

Following yet further on the possibility of the ether as a portal into higher dimensions, Karl Pearson proposed in 1892 that atoms are not vortex rings, but rather merely points where an underlying four-dimensional etherlike field literally leaked out into our three-dimensional space. This "aether squirt" theory became quite popular for some time.

Hinton's oft-stated, utter conviction that a fourth dimension was an essential part of our being was not unique. The Russian self-taught journalist, philosopher, and mystic Peter Ouspensky wrote an opus entitled *Tertium Organum* (1912) in which he stated this premise even more strongly: "And when we shall see or feel ourselves in the world of four dimensions we shall see that the world of three dimensions does not really exist, and has never existed; that it was the creation of our own fantasy, a phantom host, an optical illusion, a delusion–anything one pleases excepting only reality."

All of these diverse notions about a fourth dimension were widely debated and culminated in a 1909 essay contest sponsored by *Scientific American* for the best "explanation of the Fourth Dimension." Of particular interest today, because of their prescient resemblance to arguments that would later become part of modern lore, were the stated possibility, à la Hinton, of multiple three-dimensional universes existing within a four-dimensional framework.

The development of special relativity ultimately did provide, in 1908, via Minkowski's work, a scientific basis for a fourth dimension, but not the spatial fourth dimension so cherished by Hinton, Abbott, Wells, and others. Nevertheless, Einstein's work did play at least an indirect role in rekindling a surge of cultural interest in extra spatial dimensions.

One of the chief instigators of this was Henri Poincaré, the French mathematician whose own work on symmetries of space and time played a role in the development of relativity. The relativity of length and time measurements that were a hallmark of the special theory somehow implied to Poincaré that *all* our sense perceptions were relative, including even our perception of the number of dimensions.

In his book *Science et Méthode* (1908), Poincaré wrote: "So, the characteristic property of space, that of having three dimensions, is . . . an internal property of human intelligence, so to speak." Like Hinton before him, Poincaré believed that the key to revealing the inner reality of extra dimensions involved breaking the bonds of our limited three-dimensional intuition. As he put it: "One who devoted his life to it could *perhaps* eventually be able to picture the fourth dimension." He was a tremendously influential intellectual figure in early-twentieth-century France, and his extended notion of what one might call philosophical relativism and the associated idea that the four-dimensional world was accessible to us had wide impact.

The ways in which science has had an impact upon our culture are fascinating, and no doubt deserve more discussion than I can provide here. Yet what we see in the adoption of concepts like four dimensions and relativity as a framework for other philosophical purposes is, I suspect, more universal. People adapt what they *perceive* are scientific ideas and apply them with their own particular prejudices. They pick and choose what resonates, and the results may ultimately bear little resemblance to the actual underlying science.

Among those who helped further popularize the French fascination with four dimensions was the journalist, editor, theater critic, and science fiction writer, Gaston de Pawlowski, whose *Journey to the Country of Four Dimensions* (1912) was first serialized in installments on the front page of the literary journal *Comoedia.*

Pawlowski's literary effort, like Wells's *Time Machine,* involved a voyage to the future. But unlike Wells, he used the fourth dimension as a plot device to reflect a time when the tyranny of scientists, with their three-dimensional science, would be replaced in a future Utopia, once the existence of four dimensions was revealed to the world. Whatever one may think of this premise, Pawlowski helped instill a notion that would be popular in France and elsewhere for generations: Namely, that a lack of the proper vocabulary, both visual and verbal, has hindered our ability to free our minds to fully appreciate the underlying reality of four dimensions. As he wrote:

> The vocabulary of our language is in fact conceived according to the given facts of three-dimensional space. Words do not exist which are capable of defining exactly the strange, new sensations that are experienced when one raises himself forever above the vulgar world. The notion of the fourth dimension opens absolutely new horizons for us.

It is precisely this excitement of freeing our minds, extending the range of our senses, and opening ourselves to new experiences that is so seductive.

Ultimately the growing call for a new vocabulary with which to explore our world resonated most strongly with visual artists, whose aesthetic is directly tied to pushing against the limits of our reality. I wrote earlier about Vincent Van Gogh freeing us in 1882 from the tyranny of color, and demonstrating exotica otherwise hidden in ordinary objects. But as strange and hauntingly pleasing as his images are to the modern eye, they nevertheless preserve the spatial relationships of all the objects they represent, which remain, in spite of their jarring colors, more or less instantly recognizable.

This was not to be true of a school of artists that comprised perhaps the most influential painters and sculptors of the twentieth century who, starting about 1910, also began to transform the very definition of art. One merely has to glance at Picasso's famous *Man with Violin* (1911–12) to realize that a new way of viewing the world was emerging. Even in his early *Les Demoiselles d'Avignon* (1907), one can see in the distorted faces the beginnings of what would become a characteristic trait of presenting different perspectives on parts of figure in the single plane of a painting.

It is said that a picture is worth a thousand words. But what if, as Pawlowski stressed, words fail completely? Charles Hinton spent much of his life attempting to teach others how to develop a visual intuition about four-dimensional space, as he believed he had himself done. Recall that the heart of his technique, which was reflected in essentially every other subsequent effort, including A. Square's, was to display different three-dimensional projections of a four-dimensional object as it is "rotated" in the fourth dimension. Just as one can color the six faces of a cube and display the different colors that result when one rotates it by ninety degrees in order to help visualize both the nature of the cube and precisely what is meant by the set of rotations in three dimensions, one might hope to build up a similar understanding of four-dimensional space by considering the different three-dimensional projections of a tesseract, for example.

The similarity between Hinton's approach to the tesseract and Picasso's approach to his models is striking. But is there more to it than a simple spatial operation? Certainly, Picasso never claimed there was. His famous statement, "I paint objects as I think them, not as I see them," was more a reflection of his protest against the confines of standard perspective than a claim to be interpreting higher dimensions. Just as Van Gogh fought against the tyranny of color, one might say that Picasso and his contemporaries Braque, Gris, Metzinger, Weber, and Duchamp were struggling to free us from the tyranny of space. Yet, at the same time the ultimate goals of the mathematicians and the artists were similar: to compel us to use our minds to liberate ourselves from the confines of our own experience.

Picasso was a product of the intellectual ferment of those heady times after the turn of the century, and this was also reflected in the cubist revolution, in which he was a leading figure. The circle of artists and writers at the Bateau Lavoir in Montmartre, where cubism had its origins, discussed many of the exciting ideas of the day, including extra dimensions. While cubism was born out of a sense of questioning of the traditional views of the world, if the existence of an extra dimension could provide validation for its attempt to extract a new, hidden reality in nature, all the better.

Certainly those authors who chose to write about cubism–notably Jean Metzinger and Guillaume Apollinaire–as well as related French literary figures like Jarry–and ultimately the artist perhaps most closely associated in the modern mind with this aspect of the movement, Marcel Duchamp, all explicitly described a relationship between cubist art and four dimensions, with the analogies being alternately poetic and explicit.

Witness Duchamp, in a later interview, discussing his motivation in creating one of his most famous pieces representing a higher-dimensional reality, *The Bride Stripped Bare by Her Bachelors, Even (the Large Glass),* created between 1915 and 1923:

> What we were interested in at the time was the fourth dimension. . . . Do you remember someone called, I think, Povolowski? He was a publisher, in the rue Bonaparte. . . . He had written some articles in a magazine popularizing the fourth dimension. . . . In any case, at the time I had tried to read things by Povolowski, who explained measurements, straight lines, curves, etc. That was working in my head while I worked, although I almost never put any calculations into the *Large Glass.* Simply, I thought of the idea of a projection, of an invisible fourth dimension, something you couldn't see with your eyes.

Notes, however, for *Large Glass* do contain substantial references to mathematical discussions of a fourth dimension, including the writings of Poincaré. While Duchamp claimed only a passing knowledge of these ideas, observations he made in these notes, such as, "Poincaré's explanation about n-dim'l continuums by means of the Dedekind cut of the n-1 continuum is not in error," demonstrate the depth of his interest in the topic.

Interestingly, in spite of his truly meticulous efforts to methodically attempt to portray projections of a fourth dimension–efforts that made him more than any other artist an explicit student of this mathematics–Duchamp later disavowed them. "It wasn't for love of science that I did this," he said. "On the contrary, it was rather in order to discredit it, mildly, lightly, unimportantly. But irony was present."

For Duchamp, then, as well as for his cubist, and literary contemporanes, reacting against a three-dimensional Euclidean world was subversive and thus attractive. I use the terms *three-dimensional* and *Euclidean* here in spite of the fact that there is nothing about the four-dimensional space-time of Minkowski or, for that matter, the four-dimensional projections of Hinton and others that is remotely non-Euclidean. These spaces are quite flat. Having to go beyond Euclid to consider a possible curvature of space is essentially never explicit, except perhaps in Duchamp's piece, *Stoppages,* and in the later distorted landscapes of Salvador Dali.

Yet, in the literature of cubism non-Euclideanism was rampant. Indeed, in one of the first essays on cubism, "Du Cubisme" (1912), by Albert Gleizes and Jean Metzinger, the authors state explicitly: "If we wished to tie the painters' space to a particular geometry, we should have to refer it to the non-Euclidean scholars."

Somehow what was occurring, one might argue, was a rebellion against perspective, one of the hallmarks of our three-dimensional world. Certainly a curvature of space, causing light rays to travel on curved paths, is one way to distort perspectives, but another is to imagine viewing many different three-dimensional perspectives simultaneously, which was the preferred method of the cubists. Duchamp, one of the most mathematically literate of the emerging school, employed both non-Euclidean themes and multiple perspectives. This ultimately allowed him to go even further in his art, becoming perhaps the first of the modern conceptual artists.

While the liberation achieved by abandoning three-dimensional perspective was intoxicating, it may have been inspired, at least for some, by an incorrect understanding of the developments in science at that time. I have no idea if Einstein, a notorious antiauthoritarian, coined the word *relativity* with malice aforethought, but the term carries a great deal of intellectual baggage, and has encouraged, and continues to encourage, the incorrect notion that it somehow does away with all absolutes, making truth itself relative and observer dependent. And if special relativity, which demonstrated that space and time are tied together into a four-dimensional space-time, had everything to do with absolutes, it also has virtually nothing to do with the non-Euclidean ideas that so fascinated many of the writers and artists of the time, who may have seemed in retrospect to have been inspired by it.

One must remember also that Einstein was not yet the household name he would become in 1919, following the observations of the bending of light from distant stars which confirmed the predictions of general relativity. There is no doubt that with the passing of time his perceived impact on his cultural contemporaries may be viewed as being more significant than it actually was. In any case, as I have argued, the facets of a fourth dimension that most fascinated artists and writers alike actually had little to do with the actual ideas contained in special relativity, but were at best culled and adapted from what they perceived the theory might contain, based on preexisting cultural fascinations.

In spite of the confusions regarding the nature of the four-dimensional universe implied by relativity, and about the relations between non-Euclidean geometry and the geometry of extra dimensions, the almost accidental prescience about these concepts in the literary and artistic worlds at the beginning of the twentieth century was remarkable. I have often found (for example, when I have in other books compared science fiction and science) that the confluence of ideas and language among different disciplines is simply due to the fact that when creative people think about similar problems, even from totally different vantage points, they sometimes come up with similar ways of approaching them.

An even more remarkable coincidence, perhaps, lies in the fact, as I shall next describe, that the first concrete scientific proposal for the existence of extra spatial dimensions arose not by generalizing the notions of the space-time of Minkowski, but rather by attempting to extend Einstein's general theory of relativity, building, as fortuitously envisaged by many of the cubist artists, a bridge between curved space and extra dimensions that has been central to the scientific pursuit of extra dimensions into the twenty-first century.

So, once again, life imitates art.

CHAPTER 8

THE FIRST HIDDEN UNIVERSE: AN EXTRA DIMENSION TO PHYSICS

We are such stuff
As dreams are made on, and our little life
Is rounded with a sleep

—William Shakespeare,
The Tempest

It is one thing for a writer to dream up a new hypothetical universe in which to stage a drama, and quite another to propose that such a universe might really exist. This requires a different kind of chutzpa—the kind that arises following a period of such great success building new pictures of reality that one becomes emboldened in one's predictions. I first experienced this kind of hubris when I was a graduate student at MIT in 1980. This was a heady era in particle physics and an exciting time to be a student. In less than a decade physicists had gone from clearly understanding only one of the four known forces in nature (i.e., electromagnetism) in a way that was consistent with quantum mechanics and relativity to understanding in detail all the known forces except for gravity.

It was easy to feel that we were witnessing the emergence of an astonishing new picture of the natural world. A year earlier, Sheldon Glashow and Steven Weinberg (two faculty members at nearby Harvard, where I took most of my graduate courses) had won the Nobel Prize (along with Abdus Salam) for their development in the 1960s–confirmed by experiments in the 1970s–of a theory that unified two of the four forces in nature: the electromagnetic and weak forces. The latter is the force that is responsible for many nuclear reactions that turn protons into neutrons and vice versa, and is an integral part of the process of "nuclear fusion" that powers the sun. Shortly after that a graduate student at Harvard, David Politzer, had discovered contemporaneously with a Princeton graduate student, Frank Wilczek, and his advisor David Gross, a key mathematical characteristic of a theory that was soon recognized to describe the third nongravitational force in nature, the so-called strong force between quarks, the fundamental building blocks of protons and neutrons. The theory in question, called quantum chromodynamics (QCD), provided predictions about the interactions between quarks that were previously unthinkable, and that were ultimately verified to be in agreement with experiment, leading to a Nobel Prize thirty years later for this trio.

Everywhere we turned, it seemed that the new tools of elementary particle physics–based on combining special relativity, quantum mechanics, and Maxwell's electromagnetism–were opening up doors. Emboldened by their success, physicists began to seriously consider whether they might soon be able to unify not just two forces in nature, but perhaps three or maybe even all four, within a single theoretical mathematical framework, the holy grail of "Grand Unification."

I will return to grand unification and its predictions later in this book, but took the liberty of jumping ahead chronologically here to present a brief contemporary perspective on how the excitement of discovery can be contagious and can breed the kind of confidence that allows one to address problems one would never have had the boldness to even consider otherwise. A comparable situation occurred in the second decade of the twentieth century, following the development of special and then general relativity by Einstein.

Remarkably, just as the discoveries by Faraday, Maxwell, Oersted, Ampère, and others about the relations between electricity and magnetism led Einstein and Minkowski to propose the existence of an underlying four-dimensional space-time continuum, and just as the mathematical form of electromagnetism provided the key that allowed the physicists mentioned above to solve the mysteries surrounding the strong and weak interactions, so, too, did electromagnetism play a central role in the first serious scientific proposal that other dimensions, beyond the four we experience, might actually exist.

This proposal, like grand unification some sixty years later, was motivated by a desire to unify the forces of nature, and, as would be true of grand unification, the specific mathematical form of electromagnetism provided the direction. However, unlike the case of grand unification, the direct trigger was the remarkable discovery by Einstein that the force we feel as gravity could instead by understood in terms of the curvature of space-time.

In retrospect, it is perhaps not surprising that the advent of general relativity led physicists to consider the possibility that extra dimensions might allow for a unification of what were then the two known forces in nature: gravity and electromagnetism. Einstein's theory implied that local observers could interpret the forces they felt as either due to gravity or the effects of acceleration, depending upon their frame of reference; similarly, Maxwell's relations between electric forces and magnetic forces also imply that observers can interpret the forces they feel as either electric or magnetically induced, depending upon their own state of motion. If gravity, then, could be interpreted as being due to an underlying local curvature of three-dimensional space, then could electromagnetism be somehow due to some other sort of underlying local curvature? And since curvature in an observable three-dimensional space resulted in gravity, could curvature in some unperceived new dimension be responsible for the extra force of electromagnetism?

The Finnish physicist Gunnar Nordström actually developed the first physical theory that incorporated an extra dimension in 1914, slightly before Einstein's fully developed general relativity appeared. His version of unification was in spirit the opposite of the approach outlined above, as he tried to derive gravity from electromagnetism, rather than vice versa. Nordström had in fact developed his own theory of gravity, which attempted to generalize special relativity, just as general relativity would successfully do several years later. In Nordström's theory, the universe was five-dimensional, with one extra spatial dimension, and Maxwell's electromagnetism was a force felt in every one of the dimensions. But if, for some reason, all the electromagnetic fields were independent of the extra spatial dimension (i.e., the fields were of a constant fixed magnitude in that extra dimension, but could vary in strength over the three spatial dimensions we are used to), then those of us sensitive to only the three dimensions in which electromagnetic fields could vary would measure not only electromagnetism, but an additional remnant of the fourth spatial dimension. That additional remnant was precisely Nordström's gravitational field.

Of course, once Einstein's general relativity was unveiled, interest in Nordström's ideas waned–especially interest by Einstein, who was known to have had a less than cordial relationship with Nordström. In fact, in all the subsequent proposals involving extradimensional unifications in physics up through the early 1980s, there is not a single reference to Nordström. Such was, I suppose, the danger of competing with Einstein, at least where gravity was concerned.

The person generally credited with introducing the idea of extra dimensions into mainstream physics was the German mathematician Theodor Kaluza, in a beautiful paper entitled "On the Unity Problem in Physics," in which he argued that searching for a unified worldview was "one of the great favorite ideas of the human spirit."

Kaluza also proposed a five-dimensional universe, with four spatial dimensions plus time. He was motivated in his efforts by an earlier proposal by Hermann Weyl to unify electromagnetism and gravity in a purely geometric manner, as Einstein had done for gravity alone. Thus, instead of considering an electromagnetic field as fundamental, Kaluza imagined only a gravitational field, described by a five-dimensional version of general relativity (i.e., his theory described the curvature of four spatial dimensions in terms of a gravitational field that operated in four spatial dimensions plus time).

The fundamental quantity that determines the nature of gravity in Einstein's general relativity is something called the *metric.* This is actually a set of quantities that tell you at any point in space exactly how physical distances between nearby points are related to any local coordinate system (e.g., x, y, and z coordinates that describe length, width, and height) that a local observer may set up. If space is flat, then the relation between physical distances and coordinates such as x, y, and z is generally simple. In two dimensions, for example, the square of the physical distance between two points separated by coordinates x and y is, as Pythagoras taught us, simply x^2+y^2

But on a curved space such as the surface of a sphere, the relation between physical distances and coordinates can get strange. If one maps out points on this surface by latitudes and longitudes, for example, as one does on Earth, then near the poles, where the longitudes draw closer together, the physical distances between them is very different than it is near the equator. Thus, on a map in which latitudes and longitudes are represented by perpendicular coordinate grids, Greenland looks huge. It turns out that all of the geometric information about the sphere is precisely encoded in the metric quantities that describe the changing relation between distances as a function of latitude and longitude, and that tell us how to find out the actual size of Greenland from the difference in longitude between one side of it and the other.

In a five-dimensional space, more quantities are needed to describe all the possible coordinates for any given point. If one does the mathematics, it turns out that there are five more quantities needed at every point to completely specify the geometry of such a five-dimensional space.

Kaluza the mathematician argued as follows: Imagine a five-dimensional space that has one dimension that is periodic, such as a circle, so that when you travel in this direction, you return to your starting point. A simple example of this in two dimensions is a cylinder. Further imagine that the other four dimensions in the five-dimensional space are just like the four dimensions of space and time that we experience. The force we feel as gravity is related to the geometry of these four dimensions, described completely by the metric quantities I described earlier. Now, imagine that all the metric quantities that describe the distances between nearby points along the four-dimensional slices of five-dimensional space do not change as you move around the circular fifth spatial dimension (as would also be the case for a cylinder in two dimensions). This is the same as saying that all metric quantities that describe the five-dimensional space (there are a total of fifteen of them at any point) are independent of this circular fifth dimension.

We who can only move around in three spatial dimensions sense gravity in a way that depends upon ten of the fifteen quantities that vary from point to point in our four-dimensional slice of this five-dimensional "cylinder." So what do the other quantities determine? Kaluza was able to show that four of the extra five quantities satisfy equations that are precisely those discovered by Maxwell to describe the electric and magnetic fields.

In this way, the two known forces in nature appeared to be unified in a beautiful and remarkable way, thereby suggesting that what we measure as electromagnetic fields might be merely a remnant of an underlying curvature in an invisible fifth dimension.

This is a truly amazing possibility that sounds almost too ideal not to be true. So, why did Einstein vacillate for almost two years before finally sponsoring its publication after receiving Kaluza's manuscript in 1919?

Well, in the first place the astute reader may have noticed that I spoke of "four of the extra five quantities" that describe the geometry of the five-dimensional universe. What about the extra quantity? It turns out that Kaluza essentially ignored it, for no good reason. If one does not do this, then it turns out that the theory one arrives at in four dimensions is not quite electromagnetism plus general relativity. There is an extra term, which changes the nature of gravity. In modern language this could be described as being due to the existence of an extra massless particle in nature, which we have not observed. We shall return to this issue later.

The other question that Kaluza's work completely begs is one that resembles that question young children are required to ask on the Jewish holiday of Passover: "Why is this day different from all other days?" In this case, one would ask: "Why is the fifth dimension different from all other dimensions?" To this, Kaluza provided no concrete answer. Such was the luxury, perhaps, of being a mathematician.

To be fair, it is worth noting that Kaluza himself introduced a fifth dimension as a purely mathematical convenience, and did not necessarily ascribe any physical significance to it. Indeed, in his analysis he was apologetic in tone, calling the decision to introduce such a possibility a "strongly alienating decision." He was driven to do so by a mathematical similarity pointed out by Weyl between the way in which electromagnetic fields could be written and the way in which a certain mathematical quantity, called a "connection tensor" or just "connection," based on the metric in a curved space, could be written. Because this quantity in our four-dimensional space-time is used to describe the effects of gravity, he was forced, as he put it, to consider an extra dimension that would allow additional pieces of the connection to be interpreted as electromagnetic fields in space-time. This would only be the case as long as the extra dimension itself was rather impotent, with all physical quantities (i.e., the metric) being independent of the circular fifth dimension.

Nevertheless, Kaluza was not immune to the seductions of mathematical beauty. He found the remarkable connection between the mathematical form of electromagnetism and general relativity too compelling to resist, and in the conclusion of his paper he wrote hopefully:

> Even in the face of all the physical and epistemological difficulties which we have seen piling up against the conception presented here, it is still hard to believe that all these relations in their virtually unsurpassed formal unity, should amount to the mere alluring play of a capricious accident. Should more than an empty formalism be found to reside behind these presumed connections, we would then face a new triumph of Einstein's general relativity, whose appropriate application to a five-dimensional world is our main concern here.

It would fall to later investigators to begin to ascribe possible physical meaning to Kaluza's fifth dimension, to attempt to explain why it might be invisible, and to explore the other possible physical consequences of this idea.

The first person to seriously take up this task was physicist Oskar Klein (the son of Sweden's first rabbi), who in 1926 independently discovered the mathematical relations earlier demonstrated by Kaluza. (Somewhat later, even Einstein himself became sufficiently enamored by the idea that in 1938 he and colleague Peter Bergmann essentially reproduced Klein's ideas in a paper that represents Einstein's own continuing search for a unified theory of all interactions.) Klein, who studied with one of the fathers of quantum mechanics, Niels Bohr, was motivated in his investigations to try to understand the underlying nature of various strange phenomena predicted in quantum theory, where particles can sometimes act like waves, and probability appears to replace certainty in physical predictions.

Indeed, the developments associated with this possible unification of electromagnetism and gravity were taking place even as the top theoretical physicists of the time were wrestling with the implications and mathematics of the emerging quantum mechanical understanding of atomic phenomena. The strange nature of atomic spectra–the discrete set of colors of light emitted by different gases as you heat them up–and the nature of the radiation emitted by so-called black bodies (i.e., objects that very nearly absorb all colors of radiation equally and thus appear black) as you heat them were considered to be much more urgent problems than the more esoteric possible unification of two theories that on their own held up remarkably well.

Between 1913 and 1918, Niels Bohr had developed the first quantum theory of atomic spectra by developing a series of rather ad hoc and unusual rules to "explain" the energy levels of hydrogen. It was not until 1925–26–coincident with Klein's work on extra dimensions–that Werner Heisenberg and Erwin Schrödinger independently developed self-consistent formulations of quantum mechanics, which also implied a host of associated "spooky" phenomena, to use a phrase of Einstein's, who never fully bought into the whole picture.

Still, having a solid mathematical formulation of the rules of quantum mechanics and having a full physical understanding of the theory are two different things. Unlike both special relativity and general relativity, where comprehending the mathematics can provide one with a more or less complete physical picture, quantum mechanics defies all classical intuition.

For example, in the quantum world, subatomic particles such as electrons behave at the same time as if they are both waves and particles. That is, while individual electrons may seem like particles, they can nevertheless do things baseballs never do, such as being partially transmitted through and reflected by objects simultaneously. While the equations of quantum mechanics are themselves completely deterministic, the results of experiments are not. Rather, the equations allow one to calculate the *probability* that an experiment will yield a certain result. In 1927 Werner Heisenberg discovered one of the most frustrating aspects of the newly developed quantum theory, which has become known as the Heisenberg uncertainty principle, and which stated that, independent of one's measuring apparatus, there were certain combinations of physical quantities, such as a particle's position and velocity, that could never be known with an accuracy beyond some fundamental limit, no matter how long or hard one tried.

Einstein was not the only one who was repelled by the thought of inherent indeterminacy in our knowledge of the physical world. Perhaps this uncertainty arose because of our lack of experimental knowledge of some "hidden variables" that, if we had access to them, would allow precise and arbitrarily accurate predictions of experimental phenomena.

Klein thus rediscovered Kaluza's five-dimensional unification scheme, but his motivation was somewhat different than Kaluza's. Klein, the student of Bohr, hoped that this higher-dimensional framework might explain the basis of weird quantum mechanical phenomena, like the uncertainty principle, which he thought might be understood somehow as being due to our experiencing only a four-dimensional projection of a five-dimensional universe. This was the scientific equivalent, in a very loose sense, of explaining weird paranormal phenomena by means of the agency of an invisible fourth dimension (the difference, of course, being that weird quantum phenomena actually have been experimentally shown to exist!).

It is also somewhat ironic that Klein's motivation for rediscovering Kaluza's model came from quantum mechanics, because this was precisely what Kaluza worried about as possibly killing the whole idea. As he somewhat poetically stated at the end of his 1919 paper: "In any case, every Ansatz (i.e., postulate) which claims universal validity is threatened by the sphinx of modern physics, quantum theory."

In any case, not only did Klein reproduce Kaluza's mathematics, but because he took the possible physical existence of a fifth dimension more seriously, as a physicist rather than a mathematician, he was able to examine the physical consequences, in particular for quantum theory and also for electromagnetism, of such a fifth dimension. He also addressed the question of why it might not be observable.

His solution, which was later reproduced by Einstein and Bergmann, was to argue that this extra dimension was curled in a small circle–so small, in fact, that it could not be probed with existing experiments. In this scenario one could imagine the four dimensions of space as follows: "Above" every single point in our visible three-dimensional space a small circle "sticks out" into the fourth dimension. If one suppresses one dimension and represents our three-dimensional universe as a plane, the extra dimension could therefore be pictured by lining up an infinite number of infinitely long soda straws side by side. At each point in the plane one could travel in a circle around the side of the soda straw lying on the plane, returning back to where one started.

In fact, this analogy of the soda straw is useful from another point of view. Seen from a distance, a straw looks as if it has no thickness–as if it were a simple one-dimensional line. However, upon closer examination, one sees that the straw is actually a cylinder: a two-dimensional object (two-dimensional because one can move up and down along the length of the straw, or travel in a perpendicular direction around the side of the straw). If the diameter of the straw was small enough–say, the size of a human hair–one might not be able to perceive its thickness in the second dimension without a microscope. If it was really small, even a microscope might not reveal this second dimension. And so it could be with our universe: An extra curled-up dimension lying above every point in space would be invisible if it was curled up on a subatomic scale.

While I have presented this example by appealing to our classical intuition, Klein's argument actually relied instead on the wavelike nature of elementary particles arising out of quantum mechanics. It is well known that waves are not significantly disturbed by obstacles that are much smaller than their wavelength. A water wave in the ocean, for example, moves around a small pebble without any problem, but a large rock will protect the water behind it from the disturbance produced by an oncoming wave.

The French physicist Louis de Broglie had shown in 1924 that quantum mechanics implied that a "wavelength" could be ascribed to every particle, that would inversely proportional to the particle's "momentum" (which in turn depends upon its mass times its velocity). The higher the momentum, the smaller the wavelength. Indeed, this is why objects that are much more massive than atoms tend to behave classically: Their quantum mechanical wavelengths are so small as to be invisible, so that these objects behave, for all intents and purposes, as if they were simply particles, like billiard balls.

In order for an experiment to probe some scale, the wavelengths of the particles that one sends in as probes–be they the elementary particles associated with electromagnetic radiation called photons, or some other particles, such as electrons–must be smaller than the scale that one wishes to explore. (Otherwise, the incoming wave will not be disturbed by the object one wishes to probe.) This in turn means that the momentum, and thus the energy imparted to our particle probes, must be larger than a certain amount.

As a result, Klein, and later Einstein and Bergmann, argued that if the radius of the fifth dimension was smaller than a certain amount, then in order to send particles into this extra dimension to even resolve it one would need more energy than was then currently available in existing experiments. Because of this property, the fifth dimension could exist, yet remain effectively invisible in all existing experiments.

At the same time as providing this physical mechanism to keep the fifth dimension phenomenologically viable, Klein argued that the existence of an extra curled-up dimension might explain why all electric charges come in integral multiples of the charge on an electron (i.e., why we have never discovered any object with a charge equal to, say, 1.33 or minus 2.4 times the charge on an electron). Every object has a charge equal to . . . –3,–2,–1,0,1,2,3 . . . times an electron's charge.

Remember that in the Kaluza theory, electromagnetism is a four-dimensional remnant of what one would, if one had five-dimensional sensibilities, feel as part of a five-dimensional gravitational field. Also remember that general relativity provides a relation between the underlying energy of objects moving through space with the curvature of space they thus produce.

All of this together implies that if some particle can move in the direction of the circular fifth dimension, it will have an impact upon the geometry of the fifth dimension. To this, Klein added one last feature of the quantum world–namely, that every particle also has a wavelike character. For particles whose motion in the fifth dimension is fast enough so that their quantum-mechanical wavelength is small enough to allow them to "fit" within it, then some familiar features of wave phemonena will take over. Now, on a vibrating string only certain wavelengths are allowed, which explains why longer strings, when plucked, produce lower notes than shorter strings. On a vibrating string, only certain harmonics can survive–waves whose wavelength has a specific relationship to the length of the string (that is, is equal to the length of the string, half the length of the string, one-third the length of the string, etc.). (For those of you who are getting excited at the mention of the word *string*, you may calm down. This has nothing to do with superstring theory, which we shall get to later.)

Now, if this held true for particles moving around the circular extra dimension, since a particle's quantum mechanical wavelength is determined by its velocity, then only particles with certain fixed velocities would be able to propagate all the way around the extra dimension. A fixed set of velocities implies a fixed set of energies associated with the particle. But since energy affects geometry in general relativity, then if this theory applies in the full five-dimensional space, it means that the geometry of the fourth spatial dimension will be affected in specific, discrete ways by the presence of such particles.

Remarkably, in the Kaluza theory the effect of this change in the geometry of the fourth spatial dimension would be measured in our three-dimensional space as the existence of an electric field. Since the energies are only allowed in discrete values, the resulting electric fields, which arise from electric charges that we would view as emanating from the location in our three-dimensional space where these particles start their voyages around the extra dimension, must also come in discrete steps. Thus, all charged particles would have electric charges that are discrete multiples of some basic charge. In this way, Klein proposed that an extra dimension could explain not only the existence of both gravity and electromagnetism, but also the nature of all charged objects we measure in our universe!

With so much going for it, one might wonder why the Kaluza-Klein theory (as it is now known) did not become the next big thing in physics in the 1920s and '30s, and why these physicists are not now household names, like Einstein. There are a number of reasons. First, it became clear in these decades that the laws of quantum mechanics developed by Schrödinger, Heisenberg, and later by Dirac and others, while weird in the extreme, were nevertheless perfectly consistent with all experiments and, moreover, were inconsistent with the existence of extra "hidden variables" that might somehow lead to the apparent probabilistic nature of the theory. Thus, there was no apparent need for extra dimensions in which to hide these variables.

More important, however, was the fact that while Klein's explanation for why the fifth dimension was hidden was ingenious, it was also clearly incomplete. Namely, why would the fourth spatial dimension curl up in a circle while the other three spatial dimensions did not? Compounding this issue was the residual problem of that one extra quantity related to the five-dimensional metric that Kaluza, and later Klein, continued to ignore. It was recognized clearly by the 1940s that this extra quantity would affect the nature of gravity, so that the residual theory in four dimensions would no longer precisely be described by general relativity.

Finally, and most important of all, perhaps, was the fact that the world of physics was continuing to undergo revolutionary changes. Starting in 1930, with the discovery of the neutron, the subatomic world began to become far more complex and interesting. In short order, antimatter was discovered, as was a then new force in nature, now known as the weak force, responsible for radioactive decays. Any unification of merely gravity and electromagnetism would thus fall far short of a complete description of nature. Consequently, the majority of the physics community–rightly, I would argue–began to concentrate on trying to understand this host of new experimental phenomena, and left speculations about unobserved extra dimensions aside. It would take almost half a century before events would once again drive physicists to reconsider the possibility of a new hidden universe lying just out of sight.

CHAPTER 9
THERE AND BACK AGAIN

The theoretical possibilities in a given case are relatively few and relatively simple. . . . Considering these tells us what is possible but does not tell us what reality is.

–Albert Einstein

As exciting as the possibility of hidden extra dimensions may have seemed in 1926, within a decade the direct experimental evidence for new phenomena in three dimensions had succeeded in redirecting the imagination of the physics community toward somewhat less esoteric pursuits, or at least more experimentally accessible ones.

The half-century following 1930 was one of the most productive periods in the history of physics in terms of changing our picture of the fundamental nature of matter and energy in the universe. This may seem a surprising claim, given the fact that the two greatest single developments in the field in the twentieth century–the development of general relativity, and the discovery of the laws of quantum mechanics–had both been essentially completed by this time.

Nevertheless if the theoretical advances made during the first three decades of the century revealed a hidden nature to space and time, the experimental work conducted over the next fifty years revealed a hidden universe of exotic particles and forces. This is not to say that stunning theoretical strides were not made. They were, and I will describe them. But in contrast with general relativity and even quantum mechanics, these developments derived directly from unexpected experimental evidence based on new technologies that opened important new windows on the universe. And each time a new window on the universe has been opened, surprises have inevitably followed.

In the last years of the 1920s the capstone achievement in the theoretical development of quantum mechanics had been the work of Paul Dirac, who discovered an equation describing the quantum mechanical behavior of an electron in a way that was, for the first time, completely consistent with the principles of special relativity.

One of the remarkable predictions of Dirac's equation was that there were always two different independent solutions that satisfied the equation, which described the behavior of electrons of a certain energy. One of these described a negatively charged particle, the electron, and one described a particle with equal mass but opposite–meaning positive–charge.

When this prediction first appeared, it caused some embarrassment, because while there was one known particle in nature with equal and opposite charge to the electron–namely, the proton–it had a mass almost two thousand times larger than that of the electron. At first Dirac thought that the positive particle that showed up in his equation might somehow represent the proton. But this interpretation clearly could not hold up under detailed scrutiny. At one point, in desperation, he appealed to another sort of hidden universe: He proposed that perhaps there were other, as of yet unobserved, places in the universe where positive and negatives were reversed.

Nevertheless, this embarrassing situation turned triumphant when, in the summer of 1932, the second great discovery of the post-1930 era was made. The experimental physicist Carl Anderson, while examining the tracks left by particles in cosmic rays, the high-energy particles that bombard the earth from space every moment of every day, discovered the tracks of a particle that appeared to have a mass identical to that of the electron, but a positive charge.

The technique he used was quite straightforward. As I have described, Oersted discovered in the nineteenth century that a charged particle will experience a force if it is moving through a magnetic field. The effect of this force will be to cause its trajectory to bend. If it is positively charged, it will bend one way, and if negatively charged, the other. Anderson used a device called a cloud chamber to observe the tracks of incoming cosmic rays. This device causes charged particles to leave a cloudlike track, much like that trailed by airplanes in the sky. By placing the chamber in a large magnetic field, Anderson could determine the charge of the incoming particles by observing the direction in which their trajectories curved. Particles such as protons will indeed curve in the opposite direction to electrons, but, because the former are two thousand times heavier, a proton tends to have far greater inertia, which means its path will tend to bend far less in a magnetic field of a fixed strength than that of a high-energy electron. In one of the photographs of his chamber, taken every fifteen seconds over the course of many days, Anderson saw a track whose curvature was identical to that of the high-energy electrons he was seeing, but the direction of its curvature was opposite. The positron, as it is now known, had been discovered!

Dirac's theory was vindicated, and Dirac stated, regarding his own timidity in believing in the existence of positrons, "My equation was smarter than I was!"

One of the related striking predictions of Dirac's theory was that relativity implied that all charged elementary particles should have "antiparticles" (as they have become known): particles with identical mass and opposite electric charge. Moreover, electrons and their antiparticles—indeed all particles and their antiparticles—should be able to annihilate each other, producing pure electromagnetic radiation as an end product. Anderson was able to show that the reverse process also occurs: Very energetic electromagnetic radiation, called a gamma ray, could convert into electron–positron pairs as it traversed matter. The annihilation of these particles and antiparticles back into gamma rays was also observed.

The fact that particles and antiparticles could be created in pairs from pure energy (i.e., radiation) completely changed our thinking about matter. This was the most obvious vindication of Einstein's famous relation $E=mc^2$. Even more importantly, perhaps, it has forever changed our thinking about empty space. The reason stems from that other crown jewel of quantum mechanics, the Heisenberg uncertainty principle.

As I have mentioned, the uncertainty principle states that there are certain combinations of physically observable quantities that cannot be measured at the same time with a combined accuracy better than some amount fixed by the laws of nature, not by an experimental apparatus. The most famous such combination involves the position of a particle and its momentum, both of which cannot be measured at exactly the same time. The more accurately you can determine a particle's position, for example, the less accurately you can measure in precisely which direction it is moving. A less-known combination involves energy and time. Here, the uncertainty principle tells us that the longer we measure something, the more accurately we can determine its total energy. Since all measurements take merely a finite time, however, there is always a residual uncertainty in the value of the energy that can be measured in any system.

Now, as Faraday and Maxwell told us, if an electron is moving through space it can act as the source of electromagnetic radiation. But what if some of this electromagnetic radiation were to spontaneously convert into an electron–positron pair? Classically we would say that this is impossible, because the electron and positron together weigh twice as much as the original electron, so unless the original electron is moving so fast that its total energy is more that three times its rest mass energy, it is impossible to end up with three particles after starting with one.

But we don't live in a classical universe. Quantum mechanics is, as I like to say, just like the White House: As long as no one can measure what is going on, anything goes! In this case, the uncertainty principle tells us that during some time interval that is short enough so that the energy uncertainty is large enough–larger, say, than twice the rest mass of the electron–we cannot say how many particles exist within a region we may be measuring.

There is a finite probability that there might be, for some short period, two extra particles present. For example, an electron–positron pair could spontaneously appear for a short time, and then these particles could annihilate, leaving just the original system. As long as the particle–antiparticle pair exists for a time short enough so that the uncertainty principle indicates that we cannot measure the violation of energy conservation implied by their brief presence, the laws of quantum mechanics and relativity together suggest that such a configuration is allowed.

This represents another complete revision in our fundamental understanding of the nature of space. According to this new picture, empty space is not empty at all, but involves a boiling, bubbling brew of these particle–antiparticle pairs popping in and out of nothingness. Here is yet another hidden universe lying just beyond our perception, and one that ultimately played a key role in motivating physicists to consider the possibility of yet more radical revisions in our picture of space and time.

Before jumping on the virtual particle bandwagon, one might wonder whether suggesting particle–antiparticle pairs popping in and out of the vacuum is really any different than fantasizing about psychics popping in and out of extra dimensions in order to untie knotted ropes and recover objects inside of boxes. In cooking, the proof is in the tasting. In physics, it is in the testing.

How can we test for the existence of unobservable particles? We do it just as we might work to uncover evidence of a crime we did not witness directly: by looking for indirect evidence. And so it turns out that while one cannot measure virtual particles directly, one can nevertheless measure their effect on processes we can both calculate and measure.

Niels Bohr's first great success in his emerging quantum mechanics was to correctly predict the spectrum of light emitted by hydrogen gas when it is heated. During this process, an electron can jump between discrete allowed orbits about a proton by absorbing or radiating electromagnetic waves that we can observe as visible light. The fixed nature of the frequencies/colors emitted by hydrogen was a mystery until Bohr proposed that electrons were somehow confined to such orbits.

It was the great success of Schrödinger and Heisenberg that they presented a self-consistent mechanics that allowed a precise calculation of these energy levels in hydrogen that agreed well with the measured frequencies of radiation emitted by hydrogen atoms. However, as measurements became more and more precise, a tiny discrepancy between the predicted energy levels and the levels inferred from observation emerged.

In other areas of science, such a small discrepancy might have been ignored. But such was the precision afforded by the new merging of quantum mechanics, relativity, and electromagnetism–a theory that became known as "quantum electrodynamics"–that this experimental anomaly presented a major challenge for theoretical physicists.

Shortly after World War II the physicists who had otherwise been occupied with developing the atomic bomb returned to their fundamental investigations of nature. At one of the most famous meetings in twentieth-century physics, held on Shelter Island off Long Island in New York, a group of young turks demonstrated that a proper accounting of the effects of virtual particles could yield the critical missing component that could resolve the aforementioned shift in energy levels between theory and experiment. This shift was by then known as the Lamb shift, after the experimentalist Willis Lamb, who first discovered it.

At the time the different mathematical methods used to calculate these effects were diverse, complex, and almost mysterious, representing the similarly diverse approaches to physics of the scientists involved, from the formal and prodigiously brilliant Julian Schwinger, to the informal and sometimes irreverent genius Richard Feynman, and independently by the quiet Sin-Itiro Tomonaga, all of whom would later share the Nobel Prize for their efforts. Nevertheless, with hindsight and after a "translation" paper published by the equally brilliant Freeman Dyson, it became clear that the different approaches all reflected the same underlying physical reality.

The central point of all these approaches was that it is incorrect to calculate the orbit of an electron around a proton as if these were the only two particles present. For if virtual particle–antiparticle pairs can spontaneously appear for short periods out of nothing, then the electric field experienced by the orbiting electrons must be affected by these virtual particles. As Feynman, and Schwinger independently, calculated, by properly taking this effect into account the agreement between the predictions of QED–the acronym by which quantum electrodynamics has become known–and the empirical observations was better than one part in a million, a result that remains the best-measured prediction in all of science.

With the recognition that empty space was anything but empty, a manifest need arose to try to explicitly understand what processes take place on the smallest scales that can be imagined, and in turn understand how these processes might affect the nature of physical reality on more familiar scales. As we shall see, this program would set into motion a simmering set of internal conflicts in physics that would ultimately drive theorists to new extremes of speculation.

Alert readers will note that I referred to the discovery of antiparticles as the "second great discovery" in the post-1930 era. The first occurred about four months earlier, in February 1932, although its origins date back to the dawn of the modern era. In 1896 the French physicist Henri Becquerel found that certain substances, such as uranium, spontaneously emit a strange new sort of radiation. Mystified, he called this radiation U-rays, although his contemporaries called them Becquerel rays. Ultimately it was shown that there were actually three different kinds of radiation given off by radioactive substances, which Lord Rutherford later creatively labeled alpha, beta, and gamma rays.

Over the next decade or so Rutherford and his student James Chadwick, as well as the Polish-French physicist Marie Curie, demonstrated that gamma rays were energetic forms of electromagnetic radiation, while beta rays were energetic electrons, and alpha rays consisted of the nucleus of helium, the second lightest element, with a weight about four times that of hydrogen.

At around this time the nature of recently discovered atomic nuclei such as helium was puzzling. Since atoms were neutral objects, the charge of an atomic nucleus is precisely equal and opposite to the total charge of the electrons orbiting it. But for some reason, nuclei weighed far more than the amount that could be accounted for if they simply contained protons, the heavy, positively charged objects Rutherford identified in 1919. In 1920 Rutherford imagined two different possibilities to account for this discrepancy: First, some of the protons in a nucleus were paired with electrons inside the nucleus, canceling their charge. Alternatively, perhaps there were new neutral particles in nature with a mass almost identical to that of the proton. Neither possibility had any real evidence in support of it, but the emerging laws of quantum mechanics began to argue strongly against the former.

By 1930, after Heisenberg and Schrödinger had completed their seminal work, it was recognized that to confine an electron within a region the size of an atomic nucleus would require an energy far greater than that which was available from the electric attraction of protons and electrons. Thus there seemed no way that one could resolve the apparent paradox of nuclear masses merely by adding proton–electron pairings to nuclei.

This left the possibility of a new neutral particle as the most likely option, and motivated by this the German physicists Walther Bothe and his student Herbert Becker began to utilize radioactivity itself to explore the atomic nucleus. In 1930 they reported that when they bombarded beryllium atoms with alpha particles emitted by a radioactive source made from the element polonium (named after her native Poland by Marie Curie), a strange new type of neutral radiation was emitted that could penetrate a brass plate several centimeters thick without slowing down. This was over twenty times farther than protons with comparable energy can penetrate. Moreover, this radiation did not efficiently knock electrons out of atoms in targets, ionizing them, as charged proton beams would do. The assumption was made that this penetrating neutral radiation was a type of gamma ray, that is, high-energy electromagnetic radiation.

A key clue to the true nature of this radiation was obtained via experiments performed by the daughter of Marie Curie, Irène Joliot-Curie, and her husband, Frédíric Joliot-Curie, although this pair actually misinterpreted the data. They placed a paraffin wax target in the path of this radiation and discovered that the radiation knocked protons out of the paraffin with a very high energy. Since a similar process was known to occur in which high-energy electromagnetic radiation impinging upon atoms could knock out electrons, Joliot-Curie and her husband interpreted this new effect as an analogous phenomenon caused by even higher energy gamma rays.

Because the proton is, however, almost two thousand times as heavy as the electron, to knock it out of an atom would require far more energy than appeared to be available in the original beryllium emission process. Rutherford and Chadwick recognized this fact, and in February 1932 Chadwick announced the result of a series of experiments he had performed analogous to those that had been performed by Joliot-Curie and her husband. By using different targets he demonstrated convincingly that the mystery particle could not be a massless photon, but instead had to have a mass almost identical to that of the proton. Chadwick had discovered the neutron, one of the major components of all matter, and in so doing he solved the mystery of what makes up the missing mass in atomic nuclei. For this achievement he was awarded the Nobel Prize in 1935. (In a happy coincidence, Joliot-Curie and her husband, who did not share the prize with Chadwick because of their misinterpretation of their data, won the chemistry prize that year for their discovery of artificial radioactivity.)

Chadwick's discovery revealed a whole new world, previously hidden, inside of every atomic nucleus. It is amazing, when you think about it, that less than seventy-five years ago the most abundant component in all matter, including the very atoms in our bodies, was unknown. Moreover, what is equally remarkable in retrospect is the fact that the consideration that led Chadwick to discover the neutron is really a principle that is taught in high school physics. It can be restated in a perhaps more intuitive ways as follows: If I want to knock the headlight out of an oncoming truck, I could choose to throw a piece of popcorn at it, but I would have to throw it much faster than I am likely to be able to in order to cause any damage. However, if I use a rock, I don't have to throw it very fast to achieve my goal. Chadwick used precisely this line of reasoning to work out the details of his experiment, and to demonstrate that knocking protons out of nuclei required a massive, rather than a massless projectile.

Perhaps more than anything else, however, Chadwick's discovery of the neutron opened a Pandora's box of new mysteries in elementary particle physics. Gone was the simple world of proton and electrons, gravity and electromagnetism. Suddenly the nuclei of atoms became complex amalgamations of protons and neutrons, held together by some unknown new force.

To make matters even stranger, it turned out that this new, supposedly elementary particle, a fundamental constituent of all matter, wasn't even stable. For if you take a neutron and isolate it within a box, it will, on average, decay within a paltry ten minutes or so!

How can it be that a particle that comprises the better part of every element but hydrogen can be so ephemeral, and yet continue to dominate the mass of everything we can see? A miracle of Einstein's famous relativistic connection between mass and energy saves the day, and as a result makes our lives possible.

For it works out that a neutron is only very slightly heavier than a proton–less than one part in a hundred heavier, to be exact. When neutron decay was first observed, the decay products included protons and electrons. Originally, in fact, Chadwick thought that a neutron might be a compound object, consisting of a tightly bound proton and electron. However, relativity makes this impossible, because when particles are bound to one another it takes energy to tear them apart. But, adding energy to something, according to the precepts of relativity, makes it heavier. Thus, a bound state of a proton and an electron would weigh slightly less than would the proton and electron if they were separated.

If this were the case, and a neutron were such a bound state and thus lighter than the sum of the proton plus electron masses, it would be energetically impossible for it to spontaneously decay into a free proton and an electron. The observation of neutron decay therefore implied that the mass of the neutron had to be larger than this sum, and subsequent careful measurements showed this to be the case, if just barely.

However, by the same reasoning as given above, when a neutron itself is bound in a nucleus, by forces that were then unknown, its mass would be less than the mass of a free, unbound neutron. It turns out, remarkably, that its mass changes by just enough so that it can no longer decay into a proton plus electron when it is inside a nucleus. Thus, neutrons inside nuclei are stable. As a result, complex nuclei can be stable, and we can exist.

Getting back to the neutron itself, if it were not a bound state of a proton and an electron, how could it decay into these products? All previous observations of natural radioactivity involved the disintegration of heavy complex nuclei into smaller nuclear components. Was the neutron therefore elementary, or wasn't it? And what new force could be responsible for converting neutrons into other particles? Suddenly the strange new world of elementary particle physics became even stranger, if such a thing was possible.

And if this wasn't bad enough, the decay of the neutron produced yet another puzzle. If a neutron spontaneously decayed into a proton and an electron, the law of conservation of energy tells us that the proton and electron would each be emitted with a fixed amount of energy, so that the total energy after the decay would equal the energy available from the rest mass of the neutron. However, when the decay of the neutron was observed, it turned out that the electrons that were emitted were measured to have not a fixed energy, but a variable energy, ranging over a continuum from zero energy of motion (i.e., an electron at rest) to carrying off the total energy available associated with the mass difference between the initial neutron and the emitted proton.

If energy was to be conserved in this strange new subatomic world, there was only one solution: *Another* particle–one that would be invisible to the detectors–had to be emitted in the neutron decay. In this case, this mystery particle and the electron could share the total available energy, with the mystery particle carrying off whatever energy might not be carried off by the electron.

The problem with this explanation, however, was that the mass difference between the neutron and the sum of the masses of the proton and electron is very, very small. This means that this hypothetical particle had to be very nearly massless. Moreover, in order to have escaped detection, the particle had to have no charge, and have essentially no other significant interactions with normal matter! The Italian physicist Enrico Fermi called this proposed particle a "neutrino," which, in Italian, means "little neutron."

It took another twenty years or so for the neutrino to finally be detected, and in the interim the subatomic particle menagerie had expanded even further. The neutrino was simply the first of the novel, exotic, and alien forms of elementary particles that appeared to exist in nature, associated with seemingly new forces. This particle also appeared to not exist as a part of the stuff that makes us up and also everything we see around us. Moreover, as we shall see, the nature of some of these new forces defied our very notions of how a commonsense universe should behave. Coming to grips with the mysterious plethora of new particles and forces would occupy much of the rest of the century and would ultimately lead to speculations that even these particles and forces may reflect only the very edge of reality.

One final observational development, which actually occurred before the other two I have described thus far, contributed to the intellectual excitement of the post-1930 world. Strictly speaking, it actually occurred in 1929, but it was in the 1930s that it was fully confirmed and that its utterly revolutionary implications began to be fully appreciated by the scientific community. This was the discovery by Edwin Hubble that the universe we live in is not, on its largest scales, eternal and unchanging.

A fascinating character, Hubble was sufficiently accomplished to have garnered lasting recognition even if he had not been an expert at self-promotion. A former high school athlete, Rhodes scholar, lawyer, and high school Spanish teacher, Hubble returned to his first love, science, when he was twenty-four. A decade later, following a stint as a major in World War I, Hubble moved to the Mount Wilson Observatory to use the new hundred-inch telescope that had just been completed there. In 1924 he made his first great discovery, which ultimately changed our picture of the universe as much as anything that had ever been seen before. Observing faint variable stars in the Andromeda nebula, as it was then called, he established that these objects existed at a distance of over one million light-years away, more than three times farther away than the most distant objects known to exist within our own galaxy. Before this time the conventional wisdom–established by the influential American astronomer Harlow Shapley, who was the first to determine the size of the Milky Way–held that our galaxy was in essence an island universe, containing all there is to see. Suddenly Hubble's discovery challenged this picture. The Andromeda nebula turned out to be a neighboring galaxy of comparable size to our own, and just one of what is now understood to be more than four hundred billion galaxies in the observable universe.

Could the universe be infinite in all directions, full of galaxies as far as the eye could see and beyond? Hubble proceeded over the next five years to attempt to classify the nature of distant galaxies, and in 1929 arrived at an unexpected conclusion that made his previous startling discovery pale by comparison. In that year he reported evidence that distant galaxies are, on average, moving away from us and that, moreover, their speed is proportional to their distance: Those twice as far away are moving away twice as fast!

One's first reaction upon hearing this is to conclude that we are therefore the center of the universe. Needless to say (as my wife reminds me daily), this is not the case. What it does imply, however, is that the space between galaxies is actually uniformly expanding in all directions. Put more simply, the universe is expanding. (To prove this to yourself, draw a square grid of dots on a piece of a paper, with the dots regularly spaced. Then draw a grid with the same number of dots but with a larger uniform spacing between them. Then, if you overlay one grid over the other, placing one of the dots in the second grid right over the corresponding dot in the first grid, you will see that from the vantage point of that dot, it looks like all the other dots are moving away from it, with those twice as far away shifting by twice the amount. This works no matter which dot you do this with.)

An expanding universe is in fact precisely one of the two possibilities allowed by Einstein's general relativity. Indeed, a frustration that Einstein first encountered after developing his theory and attempting to apply it to the nature of the universe as a whole was that it did not allow for a static universe unless that universe was devoid of matter. He tried to get around this problem by introducing an extra ad hoc element into his equations–called the "cosmological term"–which he thought could allow for a static solution with matter. The effect of the cosmological term was to add a small repulsive force throughout space that Einstein thought could counteract gravity on large scales, holding distant objects apart.

Unfortunately, however, he blundered. His static solution with a cosmological constant was not stable. Had Einstein had more courage of his convictions in 1916, he might have predicted either an expanding universe or a collapsing one, because these are the only two options allowed by general relativity. Once Hubble had discovered our cosmic expansion, Einstein was overjoyed and even went to visit him at Mount Wilson in 1931 so he could look through the famous telescope himself. George Gamow, physicist and author, later said Einstein confided to him that he thought his introduction of a cosmological term into his equations was his "biggest blunder." As we shall later see, being willing to discard this term immediately after it seemed unnecessary may have been yet a bigger blunder.

In any case, Hubble's discovery of cosmic expansion changed everything about the way we think of "universal" history. If the universe was now expanding, it was once smaller. Assuming the expansion has been continuous following its history backward meant that ultimately all objects in our visible universe would have been located at a single point at a finite time in the past. This implied, first of all, that our universe had a beginning. Indeed, when Hubble initially used his measured expansion rate to determine the age of the universe, he found an upper limit of two billion years. This was embarrassing, because the earth was, and is, known to be older than that, except by school boards in Ohio, Georgia, and Kansas perhaps. Fortunately, Hubble's original measurement was actually off by almost a factor of ten, establishing a now noble tradition in cosmology. With current and thankfully more precise measurements of its expansion history, we now know that the age of the universe is about fourteen billion years.

But a finite age for the universe was not the only startling implication of the observed Hubble expansion. As we continue to move back in time, the size of the region occupied by the presently observable universes decreases. Originally, macroscopic bodies such as stars and galaxies would have been crowded together in a volume smaller than the size of an atom. In this case, the physics that would have governed the earliest moments of what has now become known as the big bang would involve processes acting on the smallest scales. On these scales the strange laws of quantum mechanics reign supreme, at least as far as we know. But, as we peer back to the very beginning itself, when all the matter in the observable universe existed together at virtually a single point, the very nature of space itself, and possibly even time as well, may have been dramatically different. Perhaps the entire universe as we know it emerged from behind the looking glass, from another dimension of sight and sound. Suddenly, faced with a possible singularity at the beginning of time, truth was stranger than fiction.

While the past remains a compelling subject, the future is usually of more practical interest. And a currently expanding universe could have one of three possible futures: Either the expansion continues unabated, or it slows down but never quite stops, or it stops and the universe recollapses. Determining which of these fates awaits the cosmos, by determining the magnitude of each of the terms in Einstein's equations for an expanding universe, became one of the principal items of business for cosmology for the rest of the twentieth century. In the 1990s we thought we finally had the answer down pat. But the universe, as it has a way of doing, surprised us. As we shall see, it turns out that empty space–not matter, and not radiation–holds the key to our future. Thus, just as trying to understand our cosmic beginnings has forced us to ponder the ultimate nature of space and time, our very future may depend upon whether there is much more to empty space than meets the eye.

These revolutions in our picture of the universe at fundamental scales, from the existence of antimatter and virtual particles, to the apparent population explosion of particles and forces, and ultimately to the dynamic nature of space itself, completely transformed the landscape of physics and affected the very questions about nature that physicists might ask. Happily, many of the confusions raised by these unexpected discoveries have been resolved, as we shall see. But not all of them have been, and in the process other puzzles have arisen that have made the preliminary thrusts of physics at the beginning of the twenty-first century bear an odd resemblance to the philosophical speculations that so inspired Poincaré, Wells, Picasso, and others at the beginning of the previous century.

CHAPTER 10
CURIOUSER AND CURIOUSER . . .

After a storm comes a calm.

–Fourteenth-century proverb

The 1950s are remembered by many to be a period of relative peace and stability, at least compared to the World War and subsequent recovery that had occupied the previous decade, and the tumultuous era that was yet to come. Memories, of course, can be deceiving, and I suspect that the families of the many thousands of Korean and U.S. soldiers killed in the Korean War, and of those who lost their lives or became trapped in Communist Hungary in 1956, may think otherwise.

Whatever one's assessment of the political situation of the time, in physics it was a period of growing but exciting confusion as the implications of the remarkable discoveries of the 1930s became manifest. Part of this excitement was generated by the availability of gargantuan tools that were part of the emergence of "big science" in late 1940s, following the mammoth Manhattan Project that led to the development of the atomic bomb and an immediate, and gruesome, end to World War II.

During this period the unprecedented power of atomic weapons raised scientists up on a pedestal. While general scientific education in the United

States did not become a priority until the crisislike reaction following the Soviet Union's launching of the *Sputnik* satellite in 1957, the public began to appreciate the possibly dramatic impact of what would otherwise be considered rather esoteric physical phenomena. The newfound knowledge of the inner workings of atomic nuclei had manifested itself in the devastation wrought by nuclear weapons. But almost as if to balance the scales, physicists also invented the transistor and, with it, solid-state electronics, exploiting the strange laws of quantum mechanics to positively revolutionize our daily lives in almost every way. Today it is hard to imagine going for even an hour without depending at some time on transistors and the technology that has been developed around them.

Even biology was benefiting from knowledge on atomic scales. X-ray crystallography was enabling scientists to piece together the atomic structure of many materials, and in April 1953, Watson and Crick discovered the remarkable double-helix structure of DNA and, with it, the very basis of life itself. Or, as they put it in the concluding sentence of the paper announcing their results, in one of the most celebrated understatements in the history of science: "It has not escaped our notice that the specific pairing we have postulated immediately suggests a possible copying mechanism for the genetic material."

The potential for the future seemed endless, limited only by our imagination. And imagination was in no short supply. But at the same time, on fundamental scales at least, nature seemed to be outpacing our ability to keep up.

The onslaught had begun slowly, as early as 1937, when once again cosmic rays produced a surprise. Recall that Carl Anderson had discovered the existence of the positron in cosmic rays in 1932 by using a cloud chamber. Shortly thereafter, in England, Patrick Blackett and his young Italian colleague Giuseppe Occhialini set out, in Blackett's charming terms, "to devise a method of making cosmic rays take their own photographs." They hooked up electronic sensors above and below a cloud chamber, which produced signals when cosmic rays passed through them. These signals were transmitted to the device that controlled the expansion of the vapor in the cloud chamber, causing the tracks to be visible. In this way, instead of expanding the cloud chamber at random, as had been done previously, and catching a cosmic ray in, on average, one out of fifty such expansions, they caught a cosmic ray in each expansion.

Using this technique, physicists could study cosmic ray properties more comprehensively, and within a few years it was observed that cosmic rays appeared to be more penetrating than one would expect based on theoretical estimates for the energy loss by electrons propagating through matter. It was natural for some–particularly experimenters, perhaps–to question whether the new quantum theory predictions of energy loss rates were, in fact, correct. Ultimately, however, the problem was demonstrated to lie elsewhere when, in 1937, two different teams of researchers (one of which included Anderson) demonstrated unambiguously that the cosmic rays being observed were not electrons, but new elementary particles, almost two hundred times heavier than the electron, and about ten times lighter than the proton and neutron. The world of elementary particles was becoming even more crowded.

Theorists, not to be outdone, pointed out that in fact one of their clan had earlier "predicted" such a particle. The soon to be famous (and infamous) U.S. physicist J. Robert Oppenheimer and his colleague Robert Serber explained that in a little-known Japanese journal, in 1935, the physicist Hideki Yukawa had proposed, by analogy to the force of electromagnetism–which operates by the exchange of electromagnetic radiation (which quantum mechanics implied could also be represented by particles, i.e., photons) between charged objects–that the strong force that must bind neutrons and protons together inside of nuclei might also operate by particle exchange. Because the nuclear force is very short range, operating over only nuclear distances, Yukawa used the Heisenberg uncertainty principle to argue that the particles responsible for transmitting this force would have to be heavy, about two hundred times the mass of the electron.

Everything seemed to be falling into place . . . except that nature would not let physicists off so easily. Experiments performed over the next decade demonstrated the somewhat strange behavior of this new particle, at the time called the "mesatron." (The term *Yukon,* after Yukawa, was briefly considered but quickly abandoned as too frivolous.) Yukawa's strong nuclear force carriers should interact strongly with nuclei, and it was therefore predicted that the negatively charged mesatron should be captured by the positively charged nuclei in matter well before it could itself decay into lighter particles such as electrons and neutrinos. By 1947 it was clear that this particle interacted millions of times less strongly than these predictions suggested it should.

Instead, it turned out that this new particle, now renamed a "muon," behaved exactly like an electron, except it was two hundred times heavier. This completely unexpected development caused the famous experimental physicist and Nobel laureate I. I. Rabi to make his now often repeated remark, "Who ordered that?" We are still wondering that today!

While 1947 brought the demise of the mesatron, it also heralded the discovery of the long sought particles proposed by Yukawa. Using a new technique involving photographic emulsions to record particle tracks–a technique that was claimed to be "so simple even a theoretician might be able to do it"–the British physicist Cecil Powell and Blackett's erstwhile collaborator Occhialini were able to go to high altitudes to search for new cosmic ray signatures.

Occhialini, who had been a mountain guide, ascended to the Pic du Midi at 2,867 meters in the French Pyrenees and exposed his film to cosmic rays high in the atmosphere. Later that year, when he and Powell examined the developed emulsions in London and Bristol, Powell remembered feeling as if they had entered a whole new world. As he later wrote; "It was as if, suddenly, we had broken into a walled orchard, where protected trees flourished and all kinds of exotic fruits had ripened in great profusion."

I have rarely read a more poignant description of the joy of scientific discovery, of seeing something absolutely new, something that no human has ever witnessed before. It is what drives individuals to scale mountains, metaphorical or literal: the hidden universe, previously unknown and unobserved, but actually present, that we all seem hardwired to crave so deeply.

Powell and his collaborator's discovery of the particles that became known as pions was not the end of the road, merely a new beginning. An even stranger discovery occurred in the same year, although it took until 1950 before it was independently confirmed. In 1947, working at Manchester, George Rochester and Cecil Butler observed two unusual events involving forked tracks in cloud chambers that appeared to be due to the decays of new particles, about five times heavier than the newly discovered pions, and half as heavy as protons. In 1950, again at Pic du Midi, using a cloud chamber carted up to this high altitude just for this purpose, Blackett's group observed similar events. The situation still remained somewhat confused until 1952, when a new refined type of cloud chamber resolved that there were actually two different types of these new sorts of particles.

What made the decay events associated with these objects so strange, literally, is that while the particles involved were indeed strongly interacting, they lived about ten million million times longer than one would estimate for unstable, strongly interacting particles. Whatever property caused them to live so long was dubbed by physicists, in an act of linguistic creativity worthy of a primary school student, "strangeness," and the mysterious entities themselves became known as "strange" particles.

Powell's cosmic ray data produced yet one more shock for the physics community, much higher on the Richter scale than even the discovery of strangeness itself. In 1949, in one of the observations associated with the discovery of strangeness, Powell noticed a strange particle, which he dubbed a tau meson, that decayed into three pions. (We now call it a kaon.) Shortly thereafter came the discovery of the theta particle, which decayed into two pions. This in itself was not especially surprising, but when careful measurements were later made, it was found that the two particles had identical masses and identical lifetimes.

Why should two such different particles be otherwise so identical? One suggestion was that they were, in fact, the same particle. However, that was impossible because the final states of the two decays behaved very differently in one crucial respect–indeed, a respect that is of great significance in the context of this book. If Lewis Carroll's Alice were to observe the three-pion outgoing particle state in her looking glass, it turns out that it would be distinguishable from the three particle state as seen in her own room, just as a left hand in her world becomes a right hand when viewed in the mirror. The three different particles arrange themselves to have a certain "handedness," just as pointing three fingers in the x, y, and z directions with your right hand produces a "right-handed coordinate system," while pointing three fingers from your left hand in these three different directions produces a coordinate system that is left handed. Try it. There is no way you can rotate one configuration into the other.

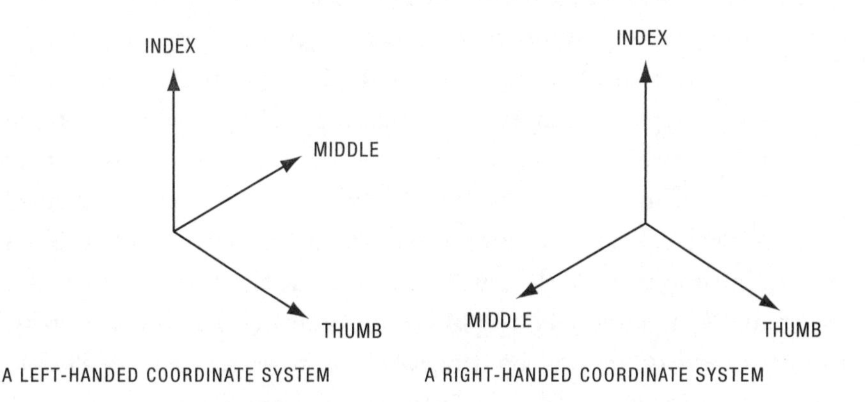

A LEFT-HANDED COORDINATE SYSTEM   A RIGHT-HANDED COORDINATE SYSTEM

By contrast, it turns out that the two-pion state would look identical in the mirror to the state as observed in the real world. There is no "handedness" to this distribution. Now, there is no way that a single particle could on the one hand produce a final state that was distinguishable from its mirror image, and on the other hand decay into a state that was identical to its mirror image, at least as long as the fundamental laws of physics governing the decays themselves don't distinguish left from right.

The tau-theta puzzle, as it became known, persisted for over five years until two young theoretical physicists, Tsung-Dao Lee and Chen Ning Yang, working for the summer at the new Brookhaven National Accelerator Laboratory in 1956, asked a remarkable question: What evidence was there that the new force responsible for the decay of these particles, the so-called weak force, which was also responsible for the decay of the neutron, actually didn't distinguish left from right?

It is hard to overstate the striking boldness of this question. After all, everything we experience about nature suggests that the world in the mirror behaves identically to our own world. Being able to distinguish left from right is simply an accident of our location. If one was out in the open ocean on a cloudy night, for example, so that one couldn't see the stars to navigate, there would be nothing on the horizon that would suggest one direction was different than any other. Or, to take a more modern example, if one was in empty space and performed any physics experiment, it would be ridiculous to expect that somehow its results should distinguish between right and left.

But there it was. In 1956 Lee and Yang realized this assumption was so ingrained in people's psyches that no one had ever bothered to test it for the weak interaction. By contrast, for both electromagnetism and for the strong interaction, this property had been verified by a host of detailed measurements. Not only did Lee and Yang recognize that no tests of left–right symmetry (or, as it has become known, *parity*) had been performed for weak interactions, they also proposed several experiments that could be performed to verify it. Within a year of the publication of their paper, two studies, both performed by physicists at nearby Columbia University, had been carried out, and both revealed the same startling conclusion: The weak interactions indeed distinguished left from right!

The first such experiment, performed by the eminent physicist Madam Chien-Shiung Wu along with collaborator Ernest Ambler at the National Bureau of Standards and his colleagues, involved nothing other than a careful observation of the decay of neutrons in the radioactive nucleus cobalt 60. Neutrons behave as if they are spinning, and if one cools down neutrons in nuclei to a very low temperature and puts them in a magnetic field, one can arrange to have most of their spin axes pointing in the same direction. When this was done for neutrons in cobalt 60, Wu and collaborators observed an angular asymmetry in the distribution of the electrons that were emitted in the decay of the neutron: More electrons were produced heading in one direction than another. With respect to the neutron spin axis, nature favored left over right. Within weeks, Leon Lederman and colleagues, also at Columbia, observed the weak decays of the recently discovered pions and muons and obtained a similar result.

Both experiments reported that the left–right asymmetry associated with weak decays was not small. Not only did nature, through weak interactions, provide a way to distinguish right from left, but it produced the maximal possible distinction. No longer could knowledgeable scientists look into the mirror and wistfully imagine a world behind the looking glass identical to their own. Like Alice, they would find that the rules in this new world were in fact different than the rules in their own world.

The astounding significance of this totally unexpected prediction of Lee and Yang's is perhaps best reflected in the fact that they were awarded the Nobel Prize in 1957, only a year, almost to the day, from the date their paper first appeared in print. Indeed, the surprise was so great that it was realized after the fact that the violation of parity had, in fact, been experimentally observed as early as 1928, even before the discovery of the neutron, in the experiments of R. T. Cox in England, who measured the scattering of electrons from the decay of radium and who detected a different scattering rate in one direction than another. His contemporaries, however, discounted his results. Sometimes, alas, it doesn't pay to be too far ahead of one's time.

The newfound complexity of the elementary particle world was both a mystery and a challenge. It also completely changed the framework for thinking about unification of forces in nature, especially along the lines of Kaluza and Klein's extra-dimensional arguments. If electromagnetism and gravity were not the only forces in nature, and if a host of new objects and strange new forces played a fundamental role, then treating electromagnetism as a residue of a purely gravitational, and thus geometric, phenomenon in higher unobserved dimensions would no longer suffice. What is surprising is that the attempt to address the mysteries brought on by these new complexities provided a completely independent impetus to consider extra dimensions.

CHAPTER 11

OUT OF CHAOS . . .

The day will perhaps come when physicists will no longer concern themselves with questions which are inaccessible to positive methods, and will leave them to the metaphysicians. That day has not yet come; man does not so easily resign himself to remaining for ever ignorant of the causes of things.

–Henri Poincaré, *Science and Hypothesis*

The startling revelations about nature discovered through cosmic rays stepped up in pace once accelerators came online, as the number and complexity of the particles produced by colliding high-energy beams on targets continued to multiply. Physics had proceeded up to that point with the presumption, generally supported by experiment, that as one probed to smaller and smaller scales the apparent complexity of the world was reduced, with increasing simplicity and economy of ideas prevailing. But this new data suggested precisely the opposite. The subatomic world appeared to be proliferating endlessly.

Two questions then naturally arose in the particle physics community: (1) Was there anything fundamental at all about any, if not all, of these particles? and (2) Would they continue to proliferate indefinitely?

By the early 1960s these concerns had given rise to several drastic proposals. One that became particularly fashionable had a certain Zen quality about it, and was for a while the dominant fad in particle theory. It was originated by physicist Geoffrey Chew at Berkeley, then the center for much of elementary particle research.

The central idea of his "bootstrap" model was that perhaps all elementary particles, and at the same time none of them, are fundamental. Put another way, perhaps all elementary particles could be viewed as being made up of appropriate combinations of other particle states. It is like imagining, for example, that combinations of the three colors red, blue, and green could make up all other colors, including themselves . . . so that red combined with blue might make green, and green combined with blue might make red. In such a case (unlike in the real world), where these three colors can be considered fundamental, the choice of which colors one considered fundamental, and which ones are composite, would clearly be arbitrary.

If you're bothered by this kind of circular thinking–oddly reminiscent of the famous "Oroboros," the snake from Indian philosophy whose head devours its own tail, ultimately disappearing completely–don't be too dismayed. Remember that the quantum mechanical world is full of apparent classical paradoxes, most of which reflect the fact that our classical notions fail to capture what are truly the essential concepts. Ultimately what the bootstrap model suggested was that perhaps particles themselves, which seem so fundamental to us, are really not the important objects to focus on, but instead are merely different reflections of some other, more basic quantities.

Perhaps instead, it was suggested the quantities that one should concentrate on were simply the mathematical relations between the different configurations that could be obtained by scattering particles off one another. The laws of quantum mechanics and relativity provide many elegant constraints on these mathematical relations, independent of the specific particles involved. Since what one actually measures in a laboratory are the processes of interactions and scattering, maybe everything that could be experimentally measured could be derived from the mathematical relationships that described the scattering of particles, and not from the classification of the properties of the particles themselves.

I am probably not doing justice to the bootstrap model, as it has since been confined to the dustbin of history. It is thus tempting to dismiss all of the work done during this period as merely a diversion, but that would not be fair. Concentration on the mathematical properties of so-called scattering amplitudes did reveal many illuminating and unexpected relations between states in the theory and the mechanisms for transformations between them.

One of the realizations that arose out of this kind of analysis was a particularly disturbing one. As more and more strongly interacting particle states were discovered, an interesting relation was discovered between the masses of particles and their "spin." Recall that many elementary particles behave as if they are spinning, and thus have an "angular momentum" similar to that of a gyroscope, which remains aligned in a certain direction and can precess about that direction and so on. The faster a top spins, the more energy it possesses, and the larger its angular momentum. Thus, it was perhaps not too surprising to find that strongly interacting elementary particles with higher-spin angular momentum tended to be heavier than their lower-spin counterparts. What was notable, however, was the roughly linear relation between the square of particle masses and their spins that was discovered.

In particular, it was tempting to predict that more and more new heavy states would be discovered as one attempted to produce states of higher and higher spin. Indeed, this prediction was verified as far out as it could be tested, so there was no reason to believe it would not carry on indefinitely.

There is a problem with this suggestion, however. If one applies the rules of quantum mechanics and relativity to calculate the scattering rates when one causes fundamental particles of higher and higher spin to collide, these rates become very large as the energy of scattering increases—much larger, in fact, than the behavior observed in actual particle-scattering experiments.

Considerations of the mathematical relations associated with scattering rates, however, offered a possible way out of this dilemma. It turned out that while the calculated rates for individual scattering processes involving the exchange of a specified number of intermediate particles of a fixed spin might grow large, it was just possible that if there were instead an infinite number of possible intermediate states and if the total scattering rate was determined by summing up over this infinite number of possibilities, then it might just be that the infinite sum could be better behaved than any of the individual terms.

I know this must sound weird in the extreme. First, how could an infinite number of particles be involved in some specific scattering process? This is made possible, however by the uncertainty principle. Remember that quantum mechanics allows for the existence of virtual particles that can spontaneously appear and disappear over short time intervals. If the interaction time is short enough, it turns out that an arbitrarily large number of virtual particles can be exchanged between the external particles undergoing a collision, with the heavier particles existing for progressively shorter times.

The second weirdness is worse, however. How could an infinite sum of terms end up being smaller than the magnitude of the individual terms in the sum? Let's warm up with a simple example. Imagine the individual terms in a sum alternate in sign–something like $1 - \frac{1}{2} + \frac{1}{3} - \frac{1}{4}$ and so on. In this case the sum of this series seems to be clearly less than 1. Namely, the sum of the first two terms is $\frac{1}{2}$, the sum of the first three is $\frac{5}{6}$, the sum of the first four is $\frac{14}{24}$, and so on. (Try adding more and more terms.)

But it turns out that infinite sums behave even more strangely. Indeed, the mathematics of infinite sums is quite fascinating and unintuitive, based as it is on the properties of infinity itself.

To get an idea about how the normal rules of addition and subtraction can become meaningless when one is considering infinite quantities, my favorite tool involves something called Hilbert's hotel, named after the famous mathematician David Hilbert, who was one of the pioneers in studying the properties of numbers, and whom I referred to earlier in the context of the development of general relativity.

Hilbert's hotel is not like a normal hotel, because it has an infinite number of rooms. Other than being rather large, you might think it would not be qualitatively different from normal hotels, but you would be wrong. For example, say that one evening Hilbert's hotel has every room occupied. In a normal hotel the manager would put up a NO VACANCY sign, but not so in this case. Say a weary traveler comes in with his family and asks for a room. The owner would happily reply that every room was now occupied, but if the traveler just waited a bit, a room would be available shortly. How would this be possible? Simple. Just take the family from room 1 and put them in room 2, the family from room 2 and put them in room 3, and so on. Since there are an infinite number of rooms, everyone gets accommodated, but now room 1 is vacant, and free for the new traveler.

Say that the new traveler arrives not merely with his family, but with his friends, as well. Because he is a very popular fellow, he brings an infinite number of friends along, each of whom wants his or her own room. No problem. The manager takes the family from room 1 and puts them in room 2, the family from room 2 and puts them in room 4, the family from room 3 and puts them in room 6, and so on. Now only the even-numbered rooms are occupied, and there are an infinite number of odd-numbered rooms vacant to accommodate the new travelers.

As these examples demonstrate, adding up infinite numbers of things is a confusing process, but mathematicians have developed rules that allow one to do so consistently. In performing such operations, however, one can find not only that the sum of an infinite series may be smaller than some of the individual terms, the sum of an infinite series can be smaller than *every single term.* Moreover, this can be the case not only for series with alternating sign terms, but for series in which every term is positive. Perhaps the most important example of this, and one of great relevance for much of the physics that follows, is the following: When considered using appropriate mathematical tools developed to handle infinite series, the sum of the series $1+2+3+4+5+\ldots$ can be shown to not equal infinity, but rather $-1/12$!

Now, in a similar vein, using similar mathematical tools, it was recognized by those who studied the mathematical relations associated with the scattering of strongly interacting particles that, if a very specific relation called "duality" (which I shall describe in more detail shortly) exists between all of the particles in the theory, then it is possible to write the total scattering rate as an infinite sum of individual contributions, each of which might blow up as the energy of the scattering particles increased, but the sum of which would instead add up to a finite number.

In 1968 the physicist Gabriele Veneziano postulated a precise formula for the scattering of strongly interacting particles that had exactly the required duality properties. It was, one should emphasize, a purely mathematical postulate, without more than at most marginal physical or experimental support. Nevertheless, the fact that it appeared to possibly resolve a conundrum that had been plaguing particle physics meant that many physicists started following up on Veneziano's ideas.

It was soon discovered that Veneziano's purely mathematical "dual model" actually did have a physical framework: through a theory not of point particles, but of "relativistic strings" (i.e., extended one-dimensional objects moving at near light-speed). Specifically, if the fundamental objects that interacted and scattered were not zero-dimensional pointlike objects, but rather one-dimensional stringlike objects, then one could show that the particular mathematical miracles associated with duality could naturally and automatically result.

Faced with the prospect of an embarrassing plethora of new particle states and also of what appeared to be an otherwise mathematically untenable theory based on that old-fashioned idea that the fundamental quantum mechanical excitations in nature are manifested as elementary particles, many physicists felt that the strong interaction had to be, at its foundation, a theory of strings.

This may all sound a bit too fantastic to be true, and those of you who are old enough to have followed popular science ideas in the 1960s and '70s may wonder why you never heard tell of strings. The answer is simple: It *was* too fantastic to be true.

Almost as soon as dual string models were developed, a number of even more embarrassing problems arose, both theoretical and experimental. The theoretical problem was, as we physicists like to say, "highly nontrivial": It turns out that when one examines the specific mathematical miracle associated with the infinite sums that duality is supposed to provide, there is a slight hitch.

The sums are supposed to produce formulae for describing the scattering of objects one measures in the laboratory. Now there is one simple rule that governs a sensible universe: If one considers all of the possible outcomes of an experiment and then conducts the experiment, one is guaranteed that one of the outcomes will actually happen. This property, which we call unitarity, really arises from the laws of probability: namely, that the sum of the probabilities of all possible outcomes of any experiment is precisely unity.

With dual string models, however, it turned out that the infinite sums in question do not, in general, respect unitarity. Put another way, they predict that sometimes when you perform an experiment, none of the allowed outcomes of the experiment will actually occur.

Thankfully, however, there turned out to be an explicit mathematical solution to this mathematical dilemma, which will be far from obvious upon first reading it, but here goes: If the fundamental objects in the theory, relativistic strings, lived not in a four-dimensional world, but a twenty-six–dimensional world, then unitarity (i.e., sensible probabilities) could be preserved.

It turns out that it is precisely the infinite sum I discussed earlier that implies this weird need for twenty-six dimensions to preserve unitarity. Considering scattering processes between strings, "virtual strings" could be exchanged, with the possibility of having an infinite number of virtual strings contributing to the scattering process. Now it turns out that the result of performing this sum yields a term that screws up the calculation of probabilities, of the following form: $[1 + \frac{1}{2}(D-2)(1+2+3+4+5+\ldots)]$, with D representing the dimension of space-time. Now, if $D=26$, and the infinite series in the second term sums up to $-\frac{1}{12}$ the total result for this offending contribution to physical scattering is precisely zero.

Now, you may recall that when Kaluza postulated the existence of a hypothetical mathematical fifth dimension, he did so sheepishly, noting "all the physical and epistemological difficulties." He essentially suggested that this extra dimension was primarily a mathematical trick, a way of unifying two disparate theories. But Kaluza's proposal was nothing compared to what appeared to be required for the consistency of dual string models—namely, that the universe must be not five-dimensional, but twenty-six-dimensional.

You might wonder whether a mathematical trick is sufficient reason to believe in twenty-two new dimensions of space, and no doubt many physicists at the time did, too. However, nature ultimately came to the rescue to resolve the debate, so that no one had to worry about this issue. Or rather, a much simpler theory than dual strings came along to completely explain the strong force.

The first inklings that dual strings might not provide the answer to the puzzling nature of the strong interaction came from experiments performed within a year or so of the time that Veneziano first proposed his mathematical solution for duality. If duality held true, then at high energies the rates of scattering of strongly interacting particles off of one another that would produce particles that flew off at a fixed angle should decline dramatically as the energy increased. But the observed falloff, while it did exist, was much less severe than the prediction.

It turned out that this finding provided clear support for an idea first proposed at the beginning of the decade by the brilliant theoretical physicist Murray Gell-Mann, who between the mid-1950s and the mid-1960s seemed to have an unerring sense of what directions might prove fruitful for unraveling the experimental confusion in elementary particle physics. Gell-Mann suggested in 1961 that one could classify the existing strongly interacting particle states into a very attractive mathematical pattern, which he called the eightfold way. What made this classification system more than mere taxonomy was that one of its first predictions is that new particles would have to exist in order to fill out some parts of the pattern that had not yet been seen. In one of the most remarkably prescient combinations of experiment and theory in recent times, in 1964 one of those new particles, called the omega-minus, was discovered, more or less exactly as Gell-Mann and his collaborators had predicted.

By 1964 Gell-Mann–and independently, George Zweig–had recognized that this underlying mathematical framework could have a physical basis if all of the dozens of strongly interacting particles, now called "hadrons," were composed of yet more fundamental particles, which Gell-Mann, the consummate scholar and linguist, dubbed "quarks" in honor of a term from James Joyce's *Ulysses*.

Quarks themselves remained a purely theoretical construct that nevertheless proved remarkably useful in classifying all the observed hadrons. However, in the late 1960s the reality of quarks as physical entities was suggested when the scattering experiments that killed the dual string picture proved instead to be completely compatible with the notion that hadrons were themselves composed of pointlike particles acting almost independently.

On its own, however, the quark model was not sufficient to explain the data. If quarks existed, why had they not been directly observed in high-energy scattering experiments? What force or forces might bind them into hadrons, and how could one explain hadron properties in terms of quark properties? And most confusing of all, if hadrons were strongly interacting, which meant that quarks had to be as well, why did the pointlike particles that appeared to make up hadrons act independently, as if they were almost noninteracting, in these high-energy scattering experiments?

Well, I already gave the punch line away several chapters earlier. In 1972–74 a series of remarkable theoretical breakthroughs basically resolved almost all the outstanding problems in elementary particle physics, as it was then understood. In particular, in a last-ditch effort to potentially put an end to what had become known as "quantum field theory," which is the theoretical framework that results when one straightforwardly combines quantum mechanics and relativity using familiar fundamental particles, David Gross at Princeton, who had been a student of Geoffrey Chew's at Berkeley during the heyday of the bootstrap model, and his own student Frank Wilczek were exploring the mathematical behavior of a type of quantum field theory called a Yang-Mills theory, named after the two physicists who had first proposed it way back in 1954.

Yang-Mills theories have another, more technical, name that is even more daunting: nonabelian gauge theories. What this term means is that these theories are similar to electromagnetism, which has a mathematical property called gauge invariance, a form of which was first explored by the mathematician Hermann Weyl in his efforts to unify electromagnetism and gravity.

An equation is said to possess a certain symmetry, or be invariant under some change, whenever that change does not alter its meaning. For example, if $A=B$, then $A+2=B+2$. Adding 2 to each side of an equation leaves the meaning of the equation invariant. If A and B represent positions in space, for example, then adding 2 to both sides of the equation would be equivalent to translating both A and B by two units in some direction. Each point would still be at the same position as the other point. This transformation is called a "translation," and the equation is said to be "translationally invariant," or possess a "translation symmetry."

Similarly, the fundamental equations of both gravity and electromagnetism remain invariant when one changes certain quantities in the theory–in the case of gravity, these include the coordinates used to measure the distance between points. As pointed out earlier the specific coordinates one uses to describe some space are chosen for convenience. The underlying physical properties, like curvature, do not depend upon the choice of coordinates. For electromagnetism however, the quantity one can freely change is related to an intrinsic characteristic of charged objects, associated, it turns out, with multiplying all charged quantities by a complex number. Weyl thought one could also make this latter quantity appear as if it were a kind of coordinate transformation, achieved by changing the scale (or "gauge") of disance measurements. One could thus "unify" the "symmetries" of electromagnetism and gravity as being associated with different kinds of coordinate transformation, but he was wrong. Nevertheless, it turns out that the separate symmetries of these two theories imply that gravity and electromagnetism share one feature in common: In both, the strength of the force between (massive or charged, respectively) objects falls off with the square of the distance between them.

It turns out that when one attempts to turn these theories into quantum theories, this particular force law, which means the force is long ranged, requires, via the uncertainty principle, the existence of a massless particle that can be exchanged between objects and by which the force is transmitted. In the case of electromagnetism this particle is called the photon, and in gravity we call the (not yet directly measured) particle the graviton.

However, in nonabelian or Yang-Mills theories, because the transformations that can leave the equations the same are more complex, instead of having only one massless force carrier field, like the photon in electromagnetism, these theories can have numerous such fields. Moreover, in electromagnetism the photon, while it is emitted and absorbed by objects that carry electric charge, does not itself carry an electric charge. But in

Yang-Mills theories the force carriers themselves are charged and thus interact with one another as well as with matter.

These theories had begun to have newfound currency in the late 1960s after it was proposed–it later turned out correctly–independently by Glashow, Weinberg, and Salam, who later shared the Nobel Prize for their insight, that one such nonabelian gauge theory could correctly describe all aspects of the weak interaction that converted protons into neutrons, and was responsible for the decay of neutrons into protons, electrons, and neutrinos.

Gross, Wilczek, and independently David Politzer, a graduate student of Sidney Coleman's at Harvard, each turned his attention to another nonabelian gauge theory whose form ultimately turned out to have certain properties that suggested it might be appropriate to describe the interactions between quarks that bound them together into hadrons.

Recall that Gross, who was trained in Chew's "bootstrap" group at Berkeley, was exploring this theory in hope of ruling it out as the last possible quantum field theory–and hence the last theory that was based on elementary particles as the fundamental quantities of interest–that might explain the exotic properties that seemed to be required to result in the high-energy scattering behavior of hadrons.

Much to his surprise, however, when he, Wilczek, and also Politzer completed their calculations, which explored precisely how virtual particles and antiparticles in this theory might affect how the force between quarks evolved as the quarks got closer together, it turned out that a miracle occurred. As Gross later put it: "For me, the discovery of asymptotic freedom was totally unexpected. Like an atheist who has just received a message from a burning bush, I became an immediate true believer."

The theory, which we now call quantum chromodynamics, or QCD for short, had precisely the property needed to explain the experimental data: Namely, the force between quarks would grow weaker as the quarks got closer–which implies, naturally, that as one pulled them farther apart the force would get stronger. This could explain why in high-energy scattering experiments the individual quarks close together inside the proton might appear almost noninteracting, while at the same time no scattering experiment had yet been successful in knocking a single quark apart from its neighbors.

Discovering the property that quark interactions grew weaker with closer proximity–which they dubbed asymptotic freedom–enabled them, and since then many other researchers, to calculate and predict very precisely the behavior of strongly interacting particles in high-energy collisions. Needless to say, the predictions have all been correct. The converse property, which suggests that the force between quarks continues to grow without bound as you try to separate them, and which has since been dubbed confinement, has not yet been fully proven to arise from QCD. However, numerical calculations with computers all suggest that it is indeed a property of the theory that is now known to describe the strong force. Gross, Wilczek, and Politzer were hence awarded the Nobel Prize in 2004 for their discovery of asymptotic freedom thirty years earlier.

Thus, out of the incredible experimental confusion of the 1940s, '50s, and '60s had ultimately arisen a beautiful set of theories, now called the standard model, that described all the known, nongravitational forces in nature in terms of rather elegant mathematical quantum field theories called gauge theories. The simplest extension of the basic laws of nature, involving quantum mechanics, relativity, and electromagnetism, had ultimately triumphed over the competing mathematical elegance of exotic ideas such as dual string models, along with their exciting, if somewhat daunting, requirement of extra dimensions.

But the game was far from over. The fatal warts of dual strings, at least as far as explaining the strong interaction, would later be turned into beauty marks in a much more ambitious program to unify gravity with the other three forces in nature. And the very properties of gauge fields and the matter that couples to them, combined with the remarkable theoretical successes that had been achieved by studying them, would lead theorists to once again revisit the very first effort to unify the first known gauge theories: gravity and electromagnetism. In so doing they would once again be driven to reconsider whether extra dimensions might be the key to understanding nature.

CHAPTER 12
ALIENS FROM OTHER DIMENSIONS

. . . the banality of existence has been so amply demonstrated, there is no need for us to discuss it any further here. The brilliant Cerebron, attacking the problem analytically, discovered three distinct kinds of dragon: the mythical, the chimeral, and the purely hypothetical. They were all, one might say, nonexistent, but each nonexisted in an entirely different way.

–Stanislaw Lem, *The Third Sally*

If physicists have been fickle in their intermittent love affair with extra dimensions, turning hot and cold as their whims and desires evolved, artists and writers have been much more faithful with their affections.

Through good times and bad, a literary fascination with another world beyond the reach of our senses has held steadfast. There is an unbroken string of writing with this focus, stretching from Lewis Carroll's *Through the Looking Glass* (1872) to C. S. Lewis's *The Lion, The Witch, and the Wardrobe* (1950) and beyond. These books, written almost a century apart, were ostensibly created for children by austere British academics, but both reach out far more broadly to that primal yearning to answer with a resounding *no,* Peggy Lee's plaintive cry: "Is that all there is?"

Despite the gap in time and intentions of the writers (Carroll was, among other things, a satirist who poked fun at both authority and then-modern mores, while C. S. Lewis wrote his tale as an allegory to promote his deep religious convictions), there is a remarkable similarity in their choice of dramatic method. Alice is transported through a looking glass to a new three-dimensional world that exists inside of the glass, but clearly not *behind* it. Lewis's Lucy similarly enters a wardrobe, which again has a well-defined back when seen from the outside, but instead of encountering a wooden frame, she stumbles into the snowy night of that other three-dimensional world, Narnia.

From a mathematical perspective (and Carroll, at least, was a mathematician), what both young girls traverse is a mystical intersection between two completely separate three-dimensional worlds. To enter one is to disappear from the other . . . or at least in Alice's case to disappear, once she turns the corner out of view of those peering into the mirror. And two separate and distinct three-dimensional worlds can intersect only if the underlying space is at least four-dimensional.

As I have previously alluded, their experience is strangely reminiscent, if less terrifying perhaps, than little Christie's experience in the *Twilight Zone* episode "Little Lost Girl." Actually, this 1962 screenplay derived from an earlier short story by Richard Matheson that appeared in the science fiction magazine *Amazing Stories* in November 1953. The contemporaneous appearance of Matheson's piece and Lewis's allegory is perhaps not surprising, for just as the world of elementary particle physics was turning topsy-turvy during the 1940s and '50s, so, too, did this period witness a resurgence of interest among writers, artists, and now filmmakers in a possible fourth spatial dimension.

During this era and the decades that followed, the extradimensional imagination of artists and writers happily moved from the purely "mythical, chimerical, and hypothetical," as per Stanislaw Lem's fanciful science fiction story, to a sensibility that was more closely attuned to emerging scientific themes. What began as a rather unrealistic fascination with the mathematical properties of a purely hypothetical fourth spatial dimension though the 1940s and '50s eventually progressed to topics like space and time travel, a host of possible new dimensions, and issues such as how information might leak in and out of our world.

Perhaps the first and best known among the modern science fiction writers who helped rekindle popular fascination with a fourth dimension was Robert Heinlein. His classic short story "And He Built a Crooked House," written in 1940 and published in the February 1941 issue of the science fiction monthly *Astounding Science Fiction,* tells the tale of an unfortunate California architect, Quintus Teal, who designs a revolutionary house based on a tesseract, which you will recall is a four-dimensional version of a cube.

Teal has a brilliant idea to save space. If you could build a tesseract house, then its footprint in our three-dimensional world could be a simple cube. But, since the full 4D tesseract has eight 3D cubical faces (as a 3D cube has six 2D square faces, you will recall), one could have an eight-room house on land with only enough space for a single room. (I understand that in later editions of *Superman* comic books, his Fortress of Solitude had a similar design, for a similar reason.)

Of course, not having access to four dimensions, Teal does the next best thing: He builds an *unfolded* tesseract. Again, just as you could unfold a cube by cutting along its edges to lay out on a piece of paper the six squares that make it up, say as follows,

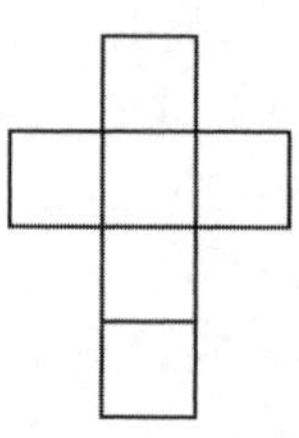

so too, you could imagine unfolding a tesseract and projecting onto a three-dimensional space the eight cubes that form its surface:

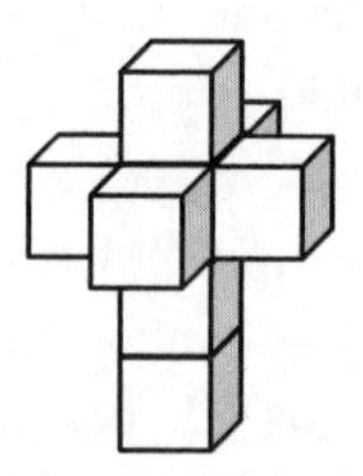

This projection, which is also called a "net," was Teal's construction—that is, until an earthquake accidentally causes the structure to fold back up into its four-dimensional form, nearly trapping Teal and the new owners in another three-dimensional space forever removed from our own.

Heinlein's fascination with hypercubes was not novel. Charles Hinton's fixation with four dimensions caused him to imagine and present a host of ways of visualizing four-dimensional objects such as tesseracts (or hypercubes, as they are also known) in all of his many writings at the turn of the last century. In the 1920s short stories continued to focus on the fourth dimension as a way to move in and out of interesting three-dimensional spaces. Both Richard Hughes's humorous "The Vanishing Man" (1926) and Miles Breuer's "The Appendix and the Spectacles" (1928) focus on the opportunities and problems that result from the fact that moving into a fourth spatial dimension would allow one to visit and explore the insides of objects, including human beings, without ever having to actually travel through their outer surfaces. But I suspect it was Heinlein's work (in particular "Crooked House") and later writing (such as Madeline L'Engle's children's classic *A Wrinkle in Time,* in which a tesseract is used as a portal to reach faraway distances and times in a folded space) that brought the idea to popular attention, and made the term *tesseract* a familiar one in popular culture. (Heinlein continued his fascination with a fourth dimension up through his 1963 story "Glory Road," which involved a hyperdimensional packing case that was bigger inside than outside.)

Coincidentally, at almost the same time as Heinlein's and Matheson's work was permeating popular culture, one of the twentieth century's most prolific and imaginative artists, Salvador Dali, who had moved well beyond cubism to help spearhead surrealism, produced his classic painting, *Crucifixion, Corpus Hypercubus,* which reproduces the tesseract net I displayed earlier.

While modern art has itself moved well beyond surrealism, so that concern with the three-dimensional notion of form has been replaced by such interests as color–or in the most extreme forms of conceptual art, no form at all–the inclusion in 1954 of a tesseract as a surrealist object of interest is part of a pattern in popular culture that I find particularly intriguing. Recall that, at the turn of the century, before, during, and immediately after the introduction of Einstein's work, fascination with a fourth spatial dimension existed entirely independently of special relativity. Yet almost a half-century after Einstein's revolutionary theories, the notion that the fourth dimension of that theory was not a spatial dimension still had not fully filtered down to the popular level. Or alternatively, even if it had, the recognition of our existence within a four-dimensional space-time continued to inspire at least a hope that other spatial dimensions might actually exist.

In particular, in both Matheson's story and Heinlein's, and in much other contemporaneous writing–such as Mark Clifton's charming short story "Star Bright" (1952) about a brilliant young girl (evolved in mental powers well beyond those of her father) who starts studying about mobius strips, Klein bottles, and tesseracts, and ultimately wills herself to step into a fourth dimension–the protagonists assume that a fourth dimension actually exists, and is common knowledge. In both "Little Lost Girl" and "Star Bright" it was implied that such a fourth dimension was a genuine concern of the physics of that period. The bewildered father in "Star Bright" exclaims: "The Moebius Strip, the Klein Bottle, the unnamed twisted cube–Einsteinian Physics. Yes, it was possible." And the physicist Bill in "Little Lost Girl" talks to the terrified couple about a portal to the fourth dimension as if it were something that everyone should be familiar with, although he does add a cautionary note: "I'm not an expert in this. . . . Who is?"

By the 1960s, however, one finds a growing and more realistic use of the intimate connection between space and time exposed by special and general relativity. Perhaps this was driven in part by the new opportunities for creative expression as special effects in movies began to blossom in the 1950s and '60s, and as television emerged as a key medium. With new graphic opportunities came new stories that exploited them. I suspect that one of the strongest driving forces, however, was the new popular fascination that began, following the 1957 launch of *Sputnik,* with the opportunities it promised of space travel.

Once attention was focused on the apparently infinite expanse of space, it was natural for scientists, and for writers and filmmakers, to speculate both about ways that one might traverse vast distances and about the large-scale nature of space itself. Stories began to proliferate in which not only does time travel become possible via transport into a fourth dimension, but also a curved space, including a curved fourth dimension, can provide spatial transport to otherwise inaccesibly distant locations.

The famous 1963 French science fiction novel by Pierre Boule, *Monkey Planet,* became an even more famous American film, *Planet of the Apes,* filmed in 1968. In the cinematic version, an astronaut who has gone on a long voyage to a distant planet later discovers that he has merely traveled in time but not in space.

Planet of the Apes makes a vague inference that somehow the long time lapse on Earth might be related to the remarkable fact, arising from special relativity, that objects moving near the speed of light relative to observers watching them have clocks that appear to be running slowly relative to the observer's clocks. This connection between space and time in relativity implies that in principle, if one was traveling at speeds close to light speed one could cross the galaxy in a single human lifetime, even if observers on the ground would measure the time elapsed for such a voyage to be many thousands of years. This fact, which (again, in principle) allows human interstellar travel without exceeding the speed of light, has become a staple of science fiction writing about space travel over the years. Indeed, as I have noted in *The Physics of Star Trek,* even the *Star Trek* writers took this fact into account, inventing a Federation "impulse drive" Speed Limit of less than half the speed of light for extended periods, so that Federation ships would not get out of time synch with their home bases.

All these applications of relativity to hypothetical space travel, in fact, involve "sub–light-speed" travel. Indeed, since special relativity suggests that travel at the speed of light is an absolute limit, there appears to be no room for anything else. However, the imaginations of writers and of scientists have offered up hopeful alternatives, of varying degrees of credibility. Almost all of these have been centered on the fact that once one allows for space itself to be dynamic–curving, expanding, or contracting in the presence of mass and energy–faster-than-light travel may be feasible. For scientists, it has been the possible dynamic evolution of space that offers the most potential. For writers, however, the curvature of space, which as I have noted is suggestive (although incorrectly so) of an embedding of our own space in a space of higher dimensions, seems to have been the primary motivating force.

The prolific Australian (British-born) science fiction writer A. Bertram Chandler was fascinated with faster-than-light travel as well as alternate realities and extra dimensions. In his story "Catch the Star Winds" (1969) he combined both ideas in a single work. The crew of the *Flying Cloud* manipulate space and time to travel faster than light and back in time, but in so doing they get hurtled into alternate dimensions from which return is impossible.

Chandler's stories may be far removed from any realistic science, but by the 1960s, physics had in fact produced some theoretical constructs associated with curved space whose properties are reminiscent of the four-dimensional objects that had earlier so fascinated Heinlein: namely, black holes and ultimately even more exotic objects called wormholes.

Black holes are remarkable not merely because they are so exotic–configurations of matter and energy so dense that the escape velocity from their surface exceeds even the speed of light–but because classically, at least, anything that falls inside one is doomed to encounter a "singularity" at their origin: A place of infinite density where the concept of space itself breaks down. Within the context of general relativity, it seems that nothing can stop the ultimate gravitational collapse of a black hole, so that the material making it up gets compressed until it achieves infinite density, at least if one follows the classical trajectories indefinitely. However, we expect that general relativity probably gets modified at small scales and high densities, where the laws of quantum mechanics hold sway.

Interestingly, however, even the classical geometry of black holes carries with it certain exotic possibilities. In particular, mathematically at least, the region inside what is called the event horizon of a black hole–the volume out of which nothing that falls in can escape–does not just extend down to the singularity, but crosses it, and connects it to another mirror image of all of space outside the event horizon. Is this just an artifact of the mathematics of classical general relativity, or could black holes act as portals to another causally disconnected universe?

Both scientists and writers have speculated about this, although if one had to pass through the singularity to get there, one probably wouldn't look too healthy coming out–rather more like meat after it has been through a grinder.

This practical issue aside, there was a compelling reason for some physicists, in particular Stephen Hawking, to have espoused the possibility that black holes are portals. When things fall through the event horizon into a black hole, one loses all subsequent information about them. Hawking showed in 1972 that when one incorporates the laws of quantum mechanics near the event horizon, black holes can actually radiate away energy, and in the process may ultimately shrink and disappear. However, it appeared mathematically that the radiation that emerged from the black hole would contain no information at all about what had fallen into it.

This is a big problem, because quantum mechanics requires that this information should be recoverable, at least in principle, if not in actual practice. If Hawking radiation really violates this principle, then quantum mechanics is definitively incompatible with gravity. Indeed, the so-called information loss paradox has been one of the central problems driving theoretical physicists to attempt to go beyond general relativity for a new theory that might be explicitly compatible with quantum mechanics.

Now, if it were really true that the information that fell down a black hole was lost to our universe, one might naturally ask where this information disappeared to. One possibility, which is the one Hawking raised, is that it would vanish down through the singularity to emerge in another universe.

Recently, however–indeed, as this book was being written–Hawking has revised his opinion. He now claims to have done a calculation that suggests that the Hawking radiation that comes out of a black hole actually does carry all the information that fell into it. If this is true, then it is a profoundly important result, as I shall later discuss, because it suggests that no significant modifications of general relativity may be necessary to resolve the information loss paradox. And, as Hawking himself has pointed out, this would also remove the prime motivation for considering black holes as portals, a fact he acknowledged with an apology: "I'm sorry to disappoint science-fiction fans."

Whether or not black holes can function as portals to another universe, the notion that there could be a potentially infinite space *inside* of objects that appear from the outside to have a finite size, extending Heinlein's "Crooked House" concept to its extreme, is actually not crazy at all.

Indeed, we may be experiencing precisely this phenomenon in the universe in which we live.

As I have discussed, our universe appears to be accelerating in its expansion, and if this acceleration is left unchecked, almost everything that astronomers can now see will recede from view, expanding infinitely far away in the infinite future. However, as Alan Guth–who in 1980 first recognized the likelihood and potential significance of periods of acceleration during the history of the universe–has demonstrated, an initially finite region of the universe that is inflating on the inside, can actually appear to be contracting when seen from the outside!

As the universe cools, certain regions can get stuck in a state that is not the true lowest-energy configuration of matter and energy, just as when one cools water down while stirring it, it can remain a liquid well below freezing. In particle physics such regions are called *false vacua.* Guth realized that a bubble of false vacuum amidst a sea of true vacuum would look very different when seen from the inside versus from the outside. Viewed from inside, the region would appear to be inflating, expanding with a constant rate of acceleration. From outside, the bubble would in fact appear to be decreasing in size, and would eventually disappear from view. Where would everything in the bubble end up? In a different, causally disconnected, and otherwise infinite universe!

This is only one of several ways that the exotic physics of curved space associated with general relativity can allow seemingly impossible things to happen. A more familiar example, perhaps, and one borrowed by Carl Sagan from physics (via his friend Kip Thorne) in his 1985 novel and then movie, *Contact,* involves wormholes. Wormholes are literally shortcuts through a curved space, much as a tunnel under a mountain saves you the travel time that would be required to cross over it. Two otherwise distant regions of space might in principle be connected via a three-dimensional wormhole if one amassed enough mass and energy at either of its mouths to produce huge local curvatures of space. However unlike a tunnel (which connects two points separated by the same linear distance apart whether or not the tunnel is there to connect them in this way), a wormhole literally changes nature of space connecting them. Before it is created there is literally no sense in which the two points might otherwise be considered close to each other.

I have written at length about the fictional fun one can have with wormholes in *The Physics of Star Trek,* so I will not repeat these discussions here. Suffice it to say that wormholes, if they actually were able to exist, would allow not only distant regions of space to be connected, but also distant regions of time . . . both past and future! But, they probably don't exist, so don't get too excited about their potential.

Star Trek, in fact, has a long history of using the effects of gravity to achieve exotic results. In one of the series's earliest episodes–and even before the physicist John Wheeler invented the term *black hole* to describe such gravitationally collapsed objects–its writers had the *Enterprise* travel too close the gravitational field of a "black star," and, as a result, the ship was thrown back in time.

While *Star Trek* has also had its share of wormholes and wormhole-induced travel, it is best known for its own faster-than-light travel mechanism, warp drive. While the very name suggests the warping of space, and I and others have discussed how, within the context of general relativity, faster-than-light travel is possible in principle (even though no information is ultimately transmitted faster than light, avoiding any contradiction with special relativity), it is interesting that in *Star Trek* lore warp drive is associated not with the dynamic warping of our own three-dimensional space, but rather with the possible existence of extra dimensions.

Indeed, while black holes and wormholes are fascinating implications of the possibility of curved space, which in some ways can mimic various phenomena one might hope would result from the existence of extra dimensions, as I have emphasized already numerous times curved space itself neither implies nor requires the existence of any extra dimensions. Instead, black holes and wormholes demonstrate that even a seemingly pedestrian three-dimensional space can be far stranger than meets the eye.

Nevertheless, as I have just described, *Star Trek* does manage to mix up warped space and extra dimensions. In fact, at the heart of the *Star Trek* universe is an apparently *infinite* number of extra dimensions. In order to explain how the crew can communicate instantly with Starfleet when they are many hundreds if not thousands of light-years away, for example, the writers invoked "subspace" communication. Using this imaginary plot device, signals are transmitted into extra subspace dimensions, where almost instantly they can be beamed back into our three-dimensional space at a different location.

Star Trek's use of subspace is characteristic of an explosion of interest beginning in the 1980s, especially in movies and television, in moving beyond merely a fourth dimension to the idea that many extra dimensions might exist, and moreover that periodically not only information but even material objects can leak from one dimension to another.

Interestingly, the things that emerge from other possible dimensions are almost never benign. A classic *Outer Limits* episode from the 1960s involved an unfortunate alien inhabitant of another dimension who meant no harm but who was accidentally sucked into our space as a result of some wayward scientific experiments, causing a host of problems. By contrast, the horror film *Poltergeist* (1980) played off the long-held notion described earlier that somehow the spirit world inhabits other dimensions that at times intersect with our own. Inevitably, however, it seems that only evil spirits choose to cross the border. In one memorable scene during the movie, reminiscent of the discussion of the magical properties of motion in higher dimensions pondered by Edwin Abbott in the nineteenth century, a ball is thrown into a dark closet and reappears by falling from the ceiling in another location of the same house.

The notion of evil beings from other dimensions plays a large part in one of my personal favorite films, *The Adventures of Buckaroo Banzai across the Eighth Dimension* (1984), whose protagonist is not only a Nobel Prize–winning particle physicist but also a skilled neurosurgeon, rock musician, and Zen warrior. Its plot focuses on the mishaps that can occur when a rocket car that can go through matter by taking a short cut through the eighth dimension picks up unwanted alien hitchhikers.

At around this time in the 1980s *Star Trek,* too, began to fixate on multiple extra spatial dimensions and the aliens that could emerge from them. In one episode Commander Riker gets kidnapped by aliens from subspace dimensions and the ship is put in great danger until the responsible portal can be closed.

Star Trek also focused on another common science fiction theme, that of parallel universes. These involve other three-dimensional spaces, identical to our own, that somehow coexist with ours, but not necessarily within the context of a higher-dimensional space. For example, some writers have taken the many-worlds interpretation of quantum mechanics, which argues that while true quantum mechanical objects can exist in many different states simultaneously, each time we make a measurement of such an object, we immediately force it to exist on one of what can be an infinite number of parallel branches of the quantum mechanical "wavefunction" describing the objects. If one takes this notion literally, it suggests that we somehow define our reality by the observations we make, but that there could be an infinite other set of realities with different outcomes.

While most physicists I know view the many-worlds interpretation as merely a mental crutch to help them deal with phenomena at the quantum level that simply have no sensible classical analogues, a number of writers have created stories using many worlds. In one *Star Trek* episode the Klingon Worf finds himself jumping between different branches of reality, in each of which all the other characters are slightly different. (Incidentally, the weird properties of quantum mechanics may have inspired artists as well as writers. More than one person I know has argued to me that Jackson Pollock's abstract "drip" paintings are beautiful representations of the quantum fluctuations that populate the vacuum.)

Up until the 1980s, the many extra dimensions proposed by science fiction and spiritual literature were essentially completely divorced from anything being considered by the scientific community. As late as 1981, for example, the idea that somehow the nature of particles and fields at the smallest scales might somehow be related to extra dimensions appeared in a story written by Craig Harrison and, in 1986, turned into a movie called *Quiet Earth.* In it, a scientist produces a fundamental change in the basic structure of matter in his laboratory, but as a result he transports almost all of the human populace to another dimension.

The nature of the confluence of extra-dimensional speculations in science and science fiction began to change, however, as notions that started to arise in elementary particle physics made their way into popular culture.

As we shall see, it was precisely the study of elementary particle physics that caused physicists to reconsider, at about the same time as Harrison's story was published, the existence of five, six, and even twenty-six dimensional spaces. And by the 1990s, after various popular accounts of the emerging research interest in the possibility of extra dimensions had appeared, one finds numerous science fiction stories–for example, "Eula Makes up Her Mind," which was featured in a recent science fiction anthology competition that I happened to judge–in which the extra dimensions of string theory play a key role. In a recent New York play the heroine somehow uses lessons from string theory in twenty-six dimensions to help her sort out her confusing love life!

As the latter example makes clear, in spite of the cross-pollination of ideas, there nevertheless remains a certain cognitive dissonance between explorations in physics and the literary allusions. I imagine that this is inevitable, and that one need not bemoan it. One of the purposes of science is to inspire people to pose questions about the universe, and if the inspiration that results is often off the mark, the effort should still be welcomed–that is, as long as people don't confuse art and reality too strongly.

Consider, after all, that from the time of Klein in the 1920s to the resurgence of interest in the topic in the 1980s and '90s, physicists were concentrating on microscopically tiny extra dimensions, so small that nothing of real interest on human scales could transpire within them or emerge from them. As I wrote in *The Physics of Star Trek,* while extra dimensions might exist, if they did, they were thought to be far too small for aliens to abduct us into them.

But, once again, life is appearing to imitate art, and to some extent science is playing catch-up. As I shall describe, possibly infinitely large extra dimensions and even parallel universes that might house everything from stars and planets to aliens have become topics that physicists now actually discuss seriously. The story of how we got to this strange place will occupy us for the rest of this book. Whether or not the current speculations about large, or even small, extra dimensions are any more firmly grounded in reality than the extra dimensions imagined by More in 1671 to house spirits or those imagined by the *Star Trek* writers from which hostile aliens might emerge, or whether instead they resemble the fictional Cerebron's analytical discovery of three different kinds of dragon–the mythical, the chimeral, and the purely hypothetical–is, of course, the million-dollar question.

CHAPTER 13
AN ENTANGLED KNOT

My soul is an entangled knot,
Upon a liquid vortex wrought
By Intellect in the Unseen residing.
And think doth like a convict sit,
With marlinspike untwisting it,
Only to find its knottiness abiding;
Since all the tools for its untying
In four-dimensional space are lying.

–James Clerk Maxwell

While the 1960s proved to be a period of discovery and confusion in elementary particle physics, as I have described, the 1970s were one of exultation, vindication, and ultimately, hubris. We began the decade mired in confusion about the quantum mechanical nature of every known force except for electromagnetism, and we completed it with a beautiful and perfectly accurate microscopic formulation of three of the four known forces in nature, with the hope of one day joining them into a single Grand Unified Theory (GUT).

It is within this historical framework that we should view the developments that have taken place since the 1970s. While the dual string theories of the late 1960s caused some physicists to take what so far appears to have been a dead-end detour to explore how microscopic extra dimensions might explain the physics of strongly interacting particles, the subsequent remarkable advances of the 1970s ultimately emboldened physicists to attempt to address the "really big" questions. Just as Einstein's great success gave him the hubris, and stamina, to devote the final thirty years of his life to an (ultimately futile) effort to produce a unified theory of all interactions, so, too, in the 1980s did physicists begin to reexamine ideas ranging from the Kaluza–Klein higher-dimensional framework to the mathematical miracles of the dual string model in an effort to once again attempt to reach Einstein's elusive goal of a unified theory.

Like all grand and ambitious campaigns, perhaps, this one began via a series of independent and sometimes serendipitous developments on seemingly unrelated fronts. These all converged in the mid-1980s in an explosion of excitement and activity that has transformed much of the focus of fundamental physics ever since.

In 1971 a young Dutch physicist, Gerardus 'tHooft, working on his PhD with his professor, Martinus Veltman, made one of those rare discoveries that changed the way physicists thought about fundamental physics. When Veltman had first met young 'tHooft, he told him to read the classic 1954 paper by Yang and Mills that proposed the now-famous Yang-Mills theories–the generalizations of electromagnetism that I wrote about earlier. While at the time the formalism proposed by Yang and Mills was essentially purely mathematical–there were no systems in nature that it could clearly describe–its elegance had raised the interest of several key theoretical physicists. One was the Nobel laureate Julian Schwinger, who around 1959 advised his graduate student Sheldon Glashow to consider how one might use these ideas to study the weak interactions, which ultimately led to Glashow's 1961 paper for which he would win the Nobel prize. Another was Veltman, who was convinced that the symmetry associated with the Yang-Mills theories was too beautiful to not be applicable to nature.

The problem with these theories was that if one tried to use them to describe physical phenomena, such as those associated with the weak interactions, then mathematical infinities appeared to result, which were not too different than those that caused physicists working on the strong interaction to first resort to the study of dual string models. Indeed, the model proposed by Glashow, and independently by Weinberg and Salam in 1967, appeared to suffer from just such infinities, so it is interesting to note that from the period 1961 to 1971 the papers that ultimately unified the weak and electromagnetic interactions were cited in the literature by physicists less than a dozen times.

However, 'tHooft, working under Veltman's guidance, discovered in 1971 that the infinities that appeared to plague the electroweak model of Glashow, Weinberg, and Salam could cleverly be removed so that the theories made mathematical sense and their predictions could be compared with experiment to arbitrarily high precision–if one had sufficient energy to do the calculations. Within two years it was understood that both the strong and weak interactions were described by Yang-Mills "nonabelian gauge theories." Three of the four forces in nature were now understood as full quantum theories. All that remained to conquer was gravity!

In the twenty years or so following Yang and Mills's work, a handful of physicists had explored the possibility that one might extend the work of Kaluza and Klein in unifying electromagnetism and gravity to the possibility of unifying gravity and Yang-Mills theories. The rationale for this was not evident, except that it was an interesting mathematical problem.

It was immediately clear that these theories, which you may recall involve more than one "photonlike" force carrier, would require a generalization to more than five dimensions. Remember that Kaluza and Klein had been able to reproduce the force of electromagnetism in four dimensions by making the photon field a part of a five-dimensional gravitational field, with the one extra dimension invisible to us.

By 1975 or so the problem had finally been worked out by various physicists, with a complete derivation by Peter Freund and collaborator Y. M. Cho. The result was what one might expect: Namely, as one could incorporate one photon by having a gravitational field in one extra (i.e., a fifth) dimension, so one could accommodate more than one "photonlike" field, as occurs in Yang-Mills theories, by adding one extra dimension for each field, and having general relativity operate in the full multidimensional space.

This model, however, did not attract much, if any, attention, for a variety of reasons. Most important was the fact that unlike Kaluza-Klein theory, a complete solution of whose equations allows three "large" and relatively flat spatial dimensions along with a compactified and thus "small" fourth spatial dimension, it turned out that the solutions of the higher-dimensional theories were not so simple.

Since the world we happen to live in is manifestly both large and three-dimensional, one might expect that the fact that these higher-dimensional unification models did not predict such a universe would kill any interest in them whatsoever. However, as I have pointed out to in another context, putting aside some mathematical ideas is like trying to put the toothpaste back in the tube after you have squirted it out. Once they are out there, they tend to take on lives of their own.

Indeed, within a year, it was recognized that if one added additional particles and forces beyond those associated with gravity in the higher-dimensional framework, one could produce the desired compactification to a large, flat, three-dimensional space and smaller extra compact dimensions. Of course, in so doing one was deviating from the spirit of Kaluza and Klein, who hoped that all the forces in nature might arise from a single gravitational field in higher dimensions. Once additional particles and fields are introduced in these extra dimensions, much of the beauty and economy of the proposal would at first seem to fade. But beauty is in the eye of the beholder, and it would turn out that there were other, equally mathematically elegant reasons to consider such additions.

For the moment though, let us return to the spirit of Kaluza and Klein and ask, if the mathematical Yang–Mills symmetries associated with the known forces in nature were to result from the geometric properties of some underlying extra-dimensional space, how many *extra* dimensions would we need to accommodate all the known forces? The answer turns out to be seven, leading to a total of eleven space-time dimensions. Thus, at the very least, the physics of the past fifty years tells us that if extra dimensions are to be the key to understanding all of the known forces in nature, there have to be a lot of them!

Eleven dimensions may seem like a lot to accommodate, but there are some good things associated with doing so. First, it is fewer than twenty-six dimensions, which is what the dual string models naively seemed to require. At the same time, it turned out that there was an independent reason to consider spaces as large as eleven dimensions in physics, coming from a consideration of the differences between the nature of matter and the matter of nature.

When we classify all the forces in nature, one fact stands out clearly: All of these forces appear to result via the exchange of virtual particles called "bosons." Recall that in quantum mechanics various properties of elementary objects, such as energy and momentum, can take on only various discrete "quantized" values. Bosons are elementary particles whose quantum angular momentum, or "spin" as we call it, comes in integer multiples of some basic fundamental value. However, when we look at matter, there is no such restriction. The basic particles that make up matter–electrons and quarks–all have spin values that are half-integer multiples of that fundamental value, and are called "fermions." Composite objects, made up of combinations of quarks, can have either half-integer or integer spin.

Now, one may wonder about this asymmetry in nature (i.e., why forces are associated with bosons, and matter is associated with both fermions and bosons). The investigation of this asymmetry took a long and convoluted trail that ultimately ended up in–you guessed it–extra dimensions. It began in 1970, when it was realized, even before they were dashed by the development of QCD, that the dual string models in twenty-six dimensions that appeared to be consistent models actually had a serious flaw. These theories predict particles called "tachyons."

Tachyons may be familiar to people who like to watch *Star Trek,* but in the real universe of physics, tachyons are bad news. As the name suggests, they have something to do with time. Strictly speaking, tachyons are particles that can appear to move backward in time, which is something that at the very least is embarrassing. Alternatively, it turns out that one can think of this behavior as due to the fact that they are particles that are restricted to always travel faster than the speed of light. Because of the relation between relative time and velocity for different observers in special relativity, it turns out that particles that somehow are forever moving faster than the speed of light (nothing can cross the threshold from slower to faster in the theory) would behave to other observers as if they are moving backward in time.

Now, it turns out that the laws of classical physics do not forbid such unusual particles to exist, but all sensible theories tend not to predict them (not to mention the fact that no tachyons have ever been observed in nature). Generally, if a theory predicts a tachyonic particle, it is usually a mathematical indication of some instability in the ground state of the theory–a reflection of the fact that one has somehow misidentified what the true stable particles are. If the instability is removed, so is the tachyon.

So, on the surface, the 1970s would seem to have been a very bad time for string theory. First, QCD came along as the correct theory of the strong interaction, and second, the dual string model appeared to be unstable, anyway. But, as has happened numerous times since, string theory has demonstrated an almost chameleon-like ability to morph into something new, its flaws transforming into virtues.

The roots of such a novel version of string theory date back to 1971, when physicists André Neveu and John Schwarz, and independently Pierre Ramond, investigated ways of allowing the incorporation of half-integer spin particles (fermions) into dual string models. Their motivation at the time was to enable these models to incorporate quarks, which by then had been demonstrated to exist inside of protons and neutrons and other strongly interacting particles. If the dual models were supposed to describe strongly interacting particles, then they would have to allow for the existence of such objects.

The mechanism for doing this is somewhat technical and may seem rather unusual on first, and probably second, glance. Normally we describe distances along a string, or any other object, in terms of regular numbers. We would say, for example: "Move 5.5 units (i.e., feet, miles, whatever) along the string." However, the mechanism that Neveu, Schwarz, and Ramond investigated did not involve using normal numbers to describe such distances along the strings but instead quantities called Grassmann variables, which obey rather strange relations. For normal numbers–say, 5 and 4–$5 \times 4 = 4 \times 5$. However, for two such Grassmann quantities, A and B, it turns out that $AB = -BA$. Moreover, since this same relation must be true for the individual quantities A and B, this means that $A^2 = -A^2 = 0$ and $B^2 = -B^2 = 0$.

I mention this not because it is particularly illuminating, but because it gives a sense of the sometimes highly nonintuitive mathematical manipulations associated with some string miracles, many of which seem unphysical, at least until one gets used to them.

In any case, one of the first important developments that occurred when fermions were added to strings using this strange mechanism is that it was realized that the critical dimension on which quantum dual string theories might make sense could be reduced from twenty-six to ten dimensions. Now, ten is close to eleven, which is the number of dimensions that pure Kaluza-Klein-type arguments seemed to favor, as I discussed earlier, but as the saying goes, close is only useful in horseshoes and hand grenades. However, this development was not the end of the story.

Once fermions were added to strings, it was realized that another remarkable bit of mathematical wizardry was possible: There could exist a brand-new symmetry that related bosons (integer spin) on the string to fermions (half-integer spin) on the string. Interestingly, it had previously been thought to be impossible to have such a symmetry interchanging bosons and fermions in one's description of nature, and in fact a theorem to this effect had been proved in 1967 by the brilliant physicist and raconteur Sidney Coleman at Harvard (who you may recall was David Politzer's supervisor) and his student Jeffrey Mandula.

It turned out, however, that by introducing those weird Grassmann quantities into the picture, one could in fact circumvent the famous Coleman-Mandula theorem and instead have such a symmetry interchanging bosons and fermions in a single physical description of the natural world. Moreover, such an extended symmetry–or "supersymmetry," as it became known–ultimately seemed to be an essential part of theories of strings that contained both fermions and bosons.

Now, interestingly enough, it wasn't until the 1970s that anyone explored the idea of applying supersymmetry beyond dual strings (i.e., two-dimensional objects moving around in ten or twenty-six dimensions) and to our good old four-dimensional universe with elementary particles such as quarks and photons. In 1974, Julius Wess and Bruno Zumino wrote a pivotal paper in which they extended the relation that had held on two-dimensional strings to our four-dimensional spacetime time consisting of fermions and bosons.

The history of supersymmetry is a somewhat convoluted one, primarily because it appeared in several different places in the literature as a mathematical idea in search of a physical application. Such ideas tend to lay dormant until circumstances arise that cause physicists to latch onto them. Once they do, there tends to be an explosion of activity, as theorists smell new opportunities like sharks smell blood.

Recall that in 1974, following the discovery a year earlier that QCD was the theory of the strong force, we appeared to have had for the first time a full quantum mechanical understanding of all the nongravitational forces in nature. Prompted by this development, Sheldon Glashow and Howard Georgi made the first proposal that same year to unify these forces in a grand unified theory.

Glashow and Georgi had written down a simple extension of the existing theories that not only appeared to unify these three nongravitational forces using a simple mathematical framework, but also nicely classified all of the known elementary particles at the same time.

On the surface, it might seem like folly to try and unify three forces whose intrinsic strengths are so different. The electric force beween quarks is tens of thousands of times less powerful than the strong force between quarks within a proton, for example. However, the beauty of asymptotic freedom was that it demonstrated that the strong force gets weaker as you measure it on smaller scales. Perhaps on some very small scale the strengths of all the forces might become similar.

Just such a calculation was first performed by Georgi and Weinberg, along with physicist Helen Quinn; it demonstrated that the quantum dynamics of the known forces was such that the difference in their strengths should indeed diminish if one examined nature on ever-smaller scales, with the strong force becoming weaker and the electromagnetic force stronger, for example. If one extrapolated to much smaller scales the known behavior at scales one could measure in the laboratory and assumed this behavior persisted without any fundamentally new physical phenomena entering in to change the results along the way, then on a scale approximately one million billion times smaller than the size of the proton, the three known forces would have approximately the same strength. What better signature of possible unification could one expect?

Everything now pointed to a simple unification of the strong, weak, and electromagnetic interactions, which, I believe, Glashow dubbed "grand unification." Moreover, this theory was not merely a convenient form of taxonomy but actually made new predictions. The boldest was that the basic building block of all matter, the proton, might not be stable, but could decay within a period of time that, while far longer than the current age of the universe, might nevertheless be measurable. A host of huge experiments was soon underway to attempt such a measurement. As Glashow put it: "Diamonds are not forever!"

As I have already described, the theoretical exuberance associated with the development of GUTs, following on the flush of success in explaining the electroweak and strong forces, was contagious. The response of the physics community followed a standard trend. Strings were largely forgotten, except by an earnest few, and there was a stampede to explore the possibilities of a new Theory of "Almost" Everything.

Suddenly physicists were boldly extrapolating known physics onto scales of energy, space, and time that had previously been unimaginable. These theories promised not just to explain the known forces, but also to answer longstanding puzzles such as how matter in the universe originated and whether matter is absolutely stable. Physicists were now seriously discussing questions associated with the earliest moments of the big bang, and experimentalists were building detectors to explore possible new phenomena on scales a million billion times smaller than the size of a proton!

Of course, following the first flush of romantic love invariably comes the recognition that the object of one's affections is not quite perfect. So it was with grand unification. As I have indicated, one of its key predictions was that the proton should not be absolutely stable, but should decay after a lifetime of about 10^{30} years. This is comfortably older than the current age of the universe (by a factor of about a hundred billion billion), so we don't have to sell our diamond rings for a loss quite yet. However, long as it is, it was within the reach of larger experiments, with tanks of thousands of tons of water containing enough protons so that one might expect, given the laws of probability, to find a few decaying each year. (With an average lifetime of 10^{30} years, this means that if one assembles 10^{30} protons in one place, on average, one will decay each year.)

Unhappily, alas, while these beautiful experiments have been launched, we have yet to witness the decay of a single proton. This failure has ruled out the GUT of Glashow and Georgi, although, as you can imagine, theorists have proposed other possibilities that still make the cut.

Another experimental problem has arisen, however, for even the simplest GUTs. Since 1974 the strength of the weak and strong forces has been measured with greater precision. Taking account of the new, more precise values, and examining theoretically what should happen as one probes on ever smaller scales, one finds that the different strengths of the three forces would not converge precisely together at a single scale, as seemed possible within the earlier, less accurate estimates.

Does this mean that grand unification is ultimately untenable? Not at all. For, even as many physicists at the time suggested, making an assumption that no new physics might enter in to change the scaling behavior of the fundamental forces as they evolve over fifteen orders of magnitude in size, from the proton size to the presumed scale of grand unification, was a remarkably conservative supposition. To come up such a vast "desert," as it became known, and to encounter no new or interesting physics, would at the very least defy a well-established historical tradition in the field.

But what could be the source of such new scaling behavior? It turned out that another problem, this time a theoretical one associated with the possible existence of grand unification, pointed the way.

The hierarchy problem, as it has become known, can be simply stated: Why are the energy (and mass and length) scales at which grand unification might occur, and the scale of the masses of the known elementary particles, so different? In another way of putting it, if grand unification indeed occurs at a scale a million billion times smaller than the size of the proton, why does nature choose to produce such a dramatic difference in scales?

Now, one perfectly good answer might simply be the same answer that parents give their children when they keep nagging them with the question, "Why?" The answer? "Because!"

Indeed, it could be just an accident of nature that we would have to live with, except that within the framework of the standard model of elementary particle physics, as it was formulated in 1974, such an accident should not happen! For, it turns out that when calculating the effects of virtual particles–the same objects that allow such good predictions for quantum electrodynamics, and also produce such nonsensical predictions for the energy of empty space–such a hierarchy would be unstable.

By unstable I mean that one can show that the virtual particles associated with the GUT can affect the measured value of some elementary particle masses at the weak scale, just as virtual particles in QED affect the magnitude of the spacing between energy levels in hydrogen atoms in a way that can be both calculated and measured. However, unlike the case in QED, where the corrections are extremely small, it turns out that the effect of virtual particles at a very high GUT-scale energy can be large enough so as to actually cause the masses of all the known particles to be raised up to this scale. The only way this can be avoided in general within the standard model would be if some very careful fine-tuning of parameters at the high-energy scale occurred, so that various large numbers would cancel out each other to high precision, leaving a remainder that might be fifteen to thirty orders of magnitude smaller. There are no known mechanisms in physics to make such cancellations occur in any natural way.

Indeed, this particular feature of the hierarchy problem is known as a "naturalness" problem. Now, as I like to say, unnatural acts probably don't seem unnatural at the time to those engaged in them. But naturalness in this sense has a well-defined meaning: It is "unnatural" to have a huge hierarchy between the masses of everyday particles and the mass scale associated with grand unification if quantum mechanical corrections to the former due to the latter might be large.

This problem has not been fully resolved, and it continues to present a tremendous challenge to theorists as they attempt to build models of reality. In fact, the vast difference in scales between the proton size and the scale at which grand unification might occur is itself dwarfed by another larger hierarchy. The predicted GUT energy scale is, in fact, several orders of magnitude smaller still than the energy scale where quantum mechanical effects in gravity should become important, and where, presumably the gravitational force might unify with the other forces. This latter scale, as I have mentioned, is called the Planck mass, and it is the ultimate bogeyman in physics. Once again, we can ask the question: Why is the Planck energy scale so vastly different than the scales of all the known elementary particles?

A glimmer of hope regarding these conundrums was elaborated by a number of authors, ultimately receiving widest impact in 1981 in a paper by Edward Witten, who had just moved to Princeton University on his way ultimately to the Institute for Advanced Study in Princeton, where he is now one the most highly regarded and accomplished mathematical physicists and string theorists in the world.

The hope appeared in the form of supersymmetry. Following 1974 a growing number of physicists began to get interested in the possible implications of space-time supersymmetry in nature beyond its utility solely for dual string models.

In order to understand the reasons for this interest, I should briefly present the key feature of supersymmetry as a symmetry of space-time. By connecting bosons and fermions, supersymmetry requires that for every boson in nature, there be a fermionic partner of exactly the same mass, electric charge, and so on.

However, in the world as we know it, this is manifestly not the case! No "superpartners," as they are called, of ordinary elementary particles have ever been seen. There is no evidence for a bosonic version of the electron, or for a fermionic version of the photon. Why on earth, then, would any physicist in her right mind suggest that such a symmetry might be appropriate to our understanding of nature?

Well, an optimistic physicist, of whom there have been many in recent years, would counter this argument by insisting that it is not that we haven't discovered all the particles predicted by supersymmetry, but rather that we have discovered precisely half of the particles! Isn't that progress?

This is not a completely facetious argument, because it turns out there are many symmetries in nature that are not manifest at first sight. For example, as I have already described, the laws of electromagnetism, which govern much of what we experience on a daily basis, do not distinguish between left and right. Yet, when I look out the window I can clearly distinguish the landscape to the left of me, where there happens to be a mountain at the moment, from the landscape to the right of me, where there doesn't happen to be one.

This is an example of what physicists call "spontaneous symmetry breaking," but it could just as justifiably be called an environmental accident. Namely, while an underlying law of nature may possess some symmetry, like left–right symmetry, that symmetry need not be manifest in the particular circumstances in which we find ourselves, such as me sitting in my office.

This may sound almost trivial, but the recognition that spontaneous symmetry breaking can occur in nature, along with an investigation of the physical implications of this possibility, have played a central role in many of the fundamental developments in a host of areas of physics over the past four decades. They certainly influenced the formulation of the electroweak theory by Glashow, Salam, and Weinberg. In that theory a fundamental symmetry relates certain facets of the weak force, and the electric force–namely, the two different forces turn out to be based in part on different mathematical realizations of a single theory. However, due to an accident of our circumstances–which, as we shall see momentarily, one can quantitatively and precisely probe–it turns out that environmental factors cause the weak force to end up looking much weaker than the electromagnetic force.

This happens because it turns out that due to differing interactions with a background field that is postulated to exist today throughout space, one of the particles that conveys the weak force (as its cousin, the massless photon, conveys the electromagnetic force) ends up behaving differently than the photon. In particular, the interactions of this "weak photon" with the background field make the "weak photon" behave like a very massive particle, almost hundred times as massive as the proton. This particle acts like a marble being dragged in the mud, while a photon is like a marble rolling on a smooth surface: The two marbles may be intrinsically identical, but they behave very differently due to the accidental circumstances in which they find themselves.

As a result, since the weak force is thus conveyed by an apparently massive particle, while the massless photon conveys the electromagnetic force, from our perspective the two forces look quite different.

This phenomenon is quite reminiscent of a much more familiar one on earth. We distinguish "north" from all the other directions because of a background magnetic field that makes our compasses point in that direction. However, if the earth had no magnetic field, there would be no such fundamental way to distinguish north from east.

In a similar vein, if supersymmetry were somehow a "broken" symmetry, then perhaps the superpartners of ordinary particles could behave differently than the particles we know. If, for example, they were much more massive–too massive, say, to have been created in current particle accelerators–then that might explain why none of them has yet been discovered.

Here one might wonder what the point is of inventing a new symmetry in nature and then coming up with a reason why it doesn't seem to apply to what we see. If this were all that were involved, the whole process would resemble intellectual masturbation. (I am motivated here perhaps by an infamous quip by Richard Feynman that physics is to mathematics as sex is to masturbation.) It is more than this, however–at least probably more than this–in part because the existence of broken supersymmetry might resolve the hierarchy problem.

One of the many interesting aspects of virtual particles is that their indirect effects on physically measurable quantities depend upon the spins of the virtual particles–that is, whether they are bosons (integral spin) or fermions (half-integral spin). Given otherwise identical fermions and bosons (i.e., masses, charges, etc.), the fermions will produce contributions identical in magnitude to the bosons, but opposite in sign.

This means that in a fully supersymmetric world, virtual particles can yield *zero* quantum mechanical corrections to physical quantities because for every boson there is a fermion of identical mass and charge, and the two sets of particles can produce equal and opposite contributions to all the processes. Thus, the effects of virtual particles at GUT scales, or at Planck scales, can disappear, so the low-energy mass scale of the particles we observe will be protected.

Of course, we do not live in a fully supersymmetric world. If supersymmetry exists, it is broken, and we would expect the fermionic partners of bosons, and vice versa, to have large masses. However, if the masses of the superpartners of ordinary matter are not too much larger than the masses of the heaviest particles we have now measured, then it is still possible for naturalness to be maintained even with a large hierarchy between the GUT scale and the scale of ordinary particles.

This is because the same virtual particle cancellation that in the fully supersymmetric world yields zero now yields an inexact cancellation. The magnitude of its inexactness will be precisely of the order of the mass difference between particles and their superpartners. If this mass difference is much smaller than the GUT scale, and is on the order of the weak scale masses, say, then virtual particle corrections will not induce masses for ordinary particles that are much larger than the weak scale. The hierarchy between the GUT scale and the weak scale then, while still uncomfortable, would at least be technically natural.

In the same year that Witten presented his argument regarding supersymmetry and the hierarchy problem, another calculation was performed that further bolstered the argument for both broken supersymmetry and grand unification. Recall that when one calculated the strengths of the three nongravitational forces as a function of scale assuming a desert between presently observed scales and the GUT scale, the strengths of the three forces would not converge together precisely at a single scale. However, if instead one assumes that a whole new set of superpartners of ordinary particles might appear with masses close to the weak scale, this would change the calculation. One then finds, given the current best-measured strengths of the three forces, a beautiful convergence together at a single GUT scale.

There are other indirect arguments that suggest that broken supersymmetry may actually be a property of nature. For example, it turns out that, in supersymmetric models, various otherwise apparently puzzling features of measured elementary particle properties can be explained. These include most importantly the strange fact that the so-called top quark (the heaviest known quark) is 175 times heavier than the proton, and almost 40 times heavier than the next heaviest quark, the bottom quark, and the fact that a predicted particle called the Higgs particle, associated with the breaking of the symmetry between the weak and electromagnetic interactions, has both thus far escaped detection but yet still could yield the quantum mechanical corrections necessary in the weak interaction to preserve agreement between theoretical predictions and observation. Finally, broken supersymmetry rather naturally predicts the existence of heavy, stable, weakly interacting particles that might make up the dark matter inferred to dominate the mass of our galaxy and all other galaxies.

But even before all of this–indeed, within a few years of the first GUT proposal and of Wess and Zumino's elucidation of the possibility of supersymmetry in our four-dimensional universe–there was another reason proposed for considering a supersymmetric universe, but this time not in four dimensions, but rather in eleven dimensions.

As I keep stressing, the development of GUTs set the stage for far more ambitious theoretical speculations about nature. Once scientists were seriously willing to consider scales a million billion times smaller than current experiments could directly measure, why not consider scales a *billion* billion times smaller? This scale is the Planck scale, where as I have mentioned one must come face to face with the problems of trying to unite gravity and quantum mechanics. Thus it was that from 1974 onward, a growing legion of physicists began to turn their attentions to this otherwise esoteric legacy of Einstein.

Recall that one of the issues that led to the development of supersymmetry, in the context of dual strings, was the realization that there appeared to be an unfortunate asymmetry in nature, wherein forces seem to be transmitted by bosons, while matter is made up of both bosons and fermions. In the context of general relativity this asymmetry is exacerbated. Namely, general relativity relates force (i.e., gravity) as a geometric quantity on the one hand, to the energy of matter on the other. Thus, force and matter are integrally related, and one might wonder if apparent distinctions between them are actually artificial.

One step in this direction was taken in 1978 by Eugene Cremmer, Bernard Julia, and Joel Scherk, who were following up on work a few years earlier exploring the possibility of "local supersymmetry," or, as it has become known, "supergravity," as a symmetry of nature. In the case of supergravity, the duality between bosons and fermions is extended to the case of the gravitational force. If one tries to model gravity as a quantum theory like electromagnetism, then the carrier of the gravitational force should be a massless particle called the graviton. It is a boson, like the photon, but instead of having spin one, it has spin two. Indeed, it would be the only known fundamental particle of spin two in nature, which is why gravity behaves so differently than the other forces.

Now, if supersymmetry is also a symmetry appropriate to gravity, then there would be a fermionic partner of the graviton, which is conventionally called the gravitino. This particle would couple to all other matter just as the graviton does, except that, being a fermion, it would be more comparable to the particles that make up ordinary matter, such as electrons and quarks, rather than the particles that carry forces, such as photons and gravitons.

Thus, in supergravity, the moment one introduces a graviton to carry the gravitational force, one also automatically must include a matter particle whose interactions are also determined purely by the gravitational force.

Cremmer, Julia, and Scherk realized that this relation between matter particles and the gravitational force in supergravity is in fact dimension dependent. In some sense one can think of this as being due to the fact that in many more dimensions, there are many more axes that a spinning particle can spin about, so that there are many more independent states in which a particle with fixed spin can be. In four dimensions, a particle of spin zero can only be in one state, a particle of spin one-half can exist in two spin states (which we often label up and down), and so on.

It turns out that in precisely eleven dimensions only a single type of supergravity theory is allowed. The mathematical relationships between particles of different spins that are determined by supersymmetry in this case is so restrictive that only one combination of particles that can include the graviton is possible if one is to achieve mathematical consistency.

In eleven dimensions the graviton (which I remind you is a boson with quantum mechanical spin value equal to 2) has 44 independent states, and the gravitino (a fermion with quantum mechanical spin value of 3/2) has 128 independent states. Since supersymmetry implies that if a graviton exists, so must its fermionic partner, this presents a problem, because the total number of fermionic states and bosonic states are not the same, as is also required by supersymmetry. Therefore, there must be eighty-four other bosonic states that can partner with the fermions, which one can think of as making up all the allowed particles of matter in this theory.

Eleven-dimensional supergravity can be thought of, therefore, as an ultimate theory, in which gravity and supersymmetry together determine all the allowed particles. Force and matter are uniquely determined.

Of course, once again the astute reader will note that in our four-dimensional universe there are many particles which have nongravitational interactions. Well, it turns out that in ten dimensions–which, as you may recall, happens to be the critical allowed number for dual strings with fermions included–gravity and supersymmetry almost completely constrain everything, but there turns out to be just enough wiggle room to have additional particles and their superpartners, which in fact can have Yang–Mills interactions.

By the early 1980s, therefore, were numerous independent reasons for serious physicists to actually consider ten or maybe eleven dimensions as real possibilities in theories that might unify gravity and the other interactions in nature. (The independent argument I mentioned earlier–that eleven dimensions might be necessary for a Kaluza-Klein theory incorporating all known forces–was actually derived much later.)

The circle was at that point almost complete; just one more ingredient was needed to close it.

Once again, we return to 1974. In that fateful year, two pioneers of dual string models, Joel Scherk and John Schwarz, realized that while these models proved a failure for describing the strong interaction, they had even greater potential. Recall that what dual string models did so well was get rid of pesky apparent infinities in the calculation of processes where particles of higher and higher spin were involved. Remember also that one of the negative features of the dual string models, besides producing incorrect predictions for scattering rates, was that they predicted a number of massless particles that had not yet been seen–in particular, a massless spin two particle.

Scherk and Schwarz argued that dual strings still might be the correct solution, but that perhaps they had been looking at the wrong problem: Maybe the apparently beautiful feature of dual strings could be combined with one of their negative features, not to describe a theory of the strong interaction, but instead to unify gravity and quantum mechanics!

After all, one of the reasons that gravity confounded all attempts to quantize it was that it involved a series of infinities in calculations because of the exchange of a massless spin two particle, the graviton. Here, string theory not only provided a possible way to remove such infinities, but also *automatically* predicted the existence of a particle with precisely the properties of a graviton. Indeed, as Richard Feynman had first demonstrated, any relativistic theory involving the exchange of a massless spin two particle could be shown to reproduce precisely Einstein's equations of general relativity.

Moreover, if dual strings were instead to be viewed as models of quantum gravity, then one more of those notorious warts in the theory could be turned into a beauty mark. Remember that dual strings require higher dimensions to make sense–either twenty-six or ten, depending upon the type of model. As applied to a theory of the strong interaction, this strained the bounds of credibility. However, as we have seen, ever since the time of Kaluza and Klein, efforts to unify gravity and other forces had focused on the possible existence of extra dimensions. In this sense, Scherk and Schwarz could claim they were following a noble tradition, rather than heading down a blind alley.

So it was that by 1981 all the independent ingredients were now in the air: GUTs, strings, supersymmetry, and a newfound desire to unify *all* the forces in nature. It would take some years, and a few more miracles, before many people other than Scherk (who sadly died 1980), Schwarz, and a few other diehards would join in the harvest, but the seeds had been planted. A growing group of physicists began to seriously believe that our four-dimensional universe really might be just the tip of a cosmic iceberg, with six or seven hidden dimensions lying, literally, just beneath the surface. The new love affair with extra dimensions had begun.

CHAPTER 14

SUPER TIMES FOR THE SUPERWORLD

If it is possible that there could be regions with other dimensions, it is very likely that God has somewhere brought them into being.

–Immanuel Kant

The year in which many in the particle physics community first experienced a "conversion" was a full decade after the apparent 1974 demise of dual strings–and, perhaps appropriately, a century after the publication of Edwin Abbott's *Flatland.* The twentieth century had brought more change in our technology and our fundamental understanding of the universe than anyone could have imagined in 1884. Yet at the beginning of the twenty-first century our fascination with hidden extra dimensions has, if anything, become even stronger, in large part because of the remarkable resurrection of an idea left for dead.

The road from Yang-Mills theories in 1954 to the proposal that strings might be a theory of gravity in the 1970s to the rise of supersymmetry in the early 1980s was, as I have described, a long and winding one. Most of all, it was not a road from which the destination was clearly visible on the horizon. Many different aspects of the problem were being explored independently by separate individuals and groups, and it was certainly not at all obvious in advance, in spite of the natural way in which gravity appeared to be embeddable in string theory, that much would amount from this effort.

First, as the saying goes, "Once bitten, twice shy." Many physicists had already seen how dual strings, the dominant fad of the late 1960s because of their great potential to resolve apparent mathematical inconsistencies of the strong interaction, had in fact been almost completely off the mark. Given this, it was understandable that they would be hesitant to embrace the theory again, even when applied in a different context.

In the second place, dual strings still suffered from embarrassing problems in 1974. While the theory might predict a graviton, it still also appeared to predict a tachyon, for example. And finally, no one had yet shown that it would produce a fully consistent quantum mechanical predictions for either gravity or other forces in nature. And of course, there was still the question of those pesky extra dimensions.

What is remarkable is that, as we have seen, piece by piece, different components seemed to fall into place to make the theory less unattractive and, at the same time, less removed from the rest of particle physics. Supersymmetry seemed to be needed once one put fermions on strings. GUTs suggested that the goal of unification itself was worth exploring, and then supersymmetry again seemed to offer the most attractive, and viable, scenarios for GUTs. Finally, applying supersymmetry to gravity seemed to once again suggest that extra dimensions might be called for.

Still, even with this growing level of attraction, string theories needed serious work to resolve their outstanding issues, which required the dedicated efforts of a small cadre of individuals, two of whom we have already encountered: Joel Scherk and John Schwarz.

John Schwarz appeared twice in the previous chapter: once associated with the effort to put fermion modes on strings, and once with the initial proposal (along with Scherk), suggesting that strings might yield a quantum theory of gravity. But his role will be even more significant in what follows. For a full decade during which much of the rest of the community was focused elsewhere, Schwarz and various collaborators—notably Joel Scherk, who tragically died before string theories truly achieved wide recognition–continued to do work on the theory, convinced that it must have something to do with nature.

I have known John Schwarz for over twenty years, and I am hard pressed to think of a time when he wasn't smiling, even when I was disagreeing with him. An indefatigably cheerful individual, John always seems to be optimistic. I believe, in fact, that his temperament has been an essential part of his ultimate success. Were this not the case, it is hard to imagine that he would have kept plugging away on what was apparently such a long shot. From 1974 until 1984 he and other string devotees labored in almost complete isolation on a model in which, frankly, almost no one other than they was interested. Without unflagging optimism they might have given up.

In any case, in 1977 string models received a big boost when Ferdinando Gliozzi, Scherk, and David Olive discovered a way to remove the tachyon from string theories in ten dimensions. Their solution appeared to involve supersymmetry as a symmetry not just on the string itself, as it had originally in fact been discovered, but throughout the full ten-dimensional space-time in which the strings moved. All of the particle states on the string involved equal numbers of fermions and bosons, a hallmark of space-time supersymmetry. Interestingly, this finding appeared well before four-dimensional supersymmetric GUT models were explored, four years later.

In 1981 John Schwarz and his collaborator Michael Green, another well-established string theorist, actually proved that Gliozzi and colleagues' construction indeed involved supersymmetry as a symmetry on the full ten-dimensional space, and not just on the string itself. String theory had officially become superstring theory.

The significance of this proof cannot be overemphasized, because with the unphysical tachyon state done away with, and with full supersymmetry in ten dimensions, a host of new and elegant mathematical techniques could then be applied to the problem of determining if the theory was fully consistent as a possible quantum theory of gravity. Within two years Green and Schwarz had their answer, and it rocked the physics world.

In 1984 they submitted a paper to the European journal *Physics Letters* in which they demonstrated that superstrings in ten dimensions could yield fermions, bosons, Yang-Mills fields, and gravitons in a way in which all nasty infinities appeared to be completely absent. It was a fully finite quantum theory that in principle had the potential to be, as it quickly became known, a Theory of Everything–the holy grail of physics ever since Einstein had first set out to unify gravity with the other forces in nature.

Suddenly all the diverse pieces that had occupied theorists over the past decade seemed to come together in a most remarkable way. Perhaps the most unexpected result was that this theory appeared to not only produce finite results instead of infinite ones when dealing with what seemed otherwise intractable physical processes, but if the ten-dimensional superstring had attached to it a sufficiently large set of Yang-Mills fields, then it turned out that it would be possible to break left–right symmetry. As you will recall, this is required if the theory is ultimately to incorporate in four dimensions the measured weak interaction–which has no such symmetry–without producing a mathematical horror called an *anomaly*. The response to these dramatic results from the particle physics community was thunderous.

The first result–the lack of infinities–was perhaps not so surprising. After all, strings had tamed infinities when they were proposed as models of the strong interaction. Recall that the mechanism of producing finite results was apparently based on a mathematical trick: An infinite sum of terms can add up to a finite number even if the individual terms appear to increase indefinitely. In the case of strings, because an infinite set of states exist with every higher energy, as vibrations of a string becomes more pronounced, the possibility of infinite sums contributing to any physical process is immediate. What was far less obvious was that the physical conditions associated with the quantum mechanics of strings would allow the infinite sums to, in fact, converge to a finite value.

In retrospect, there is a more concrete way of understanding this particular string miracle. Remember that quantum mechanics and relativity tell us that forces between particles occur via the exchange of virtual particles–those objects that can appear momentarily and then disappear so quickly that they cannot be directly observed. In this case, a virtual particle can be emitted by one object and absorbed by the other on an exceedingly small timescale.

Now, the troublesome mathematical infinities arise when virtual particles of arbitrarily high energy are exchanged. Because the uncertainty principle tells us that if virtual particles carry a great deal of energy, they can exist for only a very short time, the particles that can emit and absorb them, respectively, must be very, very close together. High-energy processes such as this are therefore really probing the nature of very short distance scales.

Strings solve this problem because on very short distance scales what we would otherwise view as elementary particles could instead be seen as excitations of strings. Below some distance scale, then, elementary particles must be treated as spread-out vibrations of a string. Thus, by changing the rules at short distances, strings provide a new limit (or "cutoff," as it is referred to by physicists), thus taming the otherwise potentially nasty short-distance, or high-energy, behavior of virtual processes involving point particles.

This kind of smoothing mechanism actually has another precedent–in this case arising not from earlier considerations of the strong interaction, but rather from the weak interaction. Before the weak and electromagnetic interactions were unified in a Yang-Mills–type theory, Enrico Fermi developed an approximate theory that could be used for calculating weak processes. While this theory was very good at low energies, it was well known that it would eventually produce nonsensical results if the energies involved got too high.

In the Fermi theory, weak interactions resulted from four different particles interacting at a single point (for example, when a neutron might decay into a proton, an electron, and an antineutrino). In the refined electroweak theory, however, it was seen that, in fact, what appeared at large distances to be four particles interacting at a single point was really two particles emitting a virtual particle that traveled a very short distance before either being absorbed by or producing via its decay, the other two particles. The short distance scale–at which this new picture becomes manifest provided a short distance cutoff in calculations. Namely, the calculations of the old theory were only valid if one considered processes on scales larger than this short distance-limiting scale. On smaller scales new rules would apply, which, in fact, turned the previously nonsensical results into finite, sensible predictions that could be compared with experiments.

This new short-distance scale where the rules change, called the "weak scale," turns out to be precisely the scale below which the particles that convey the weak force behave differently than photons, which convey the electromagnetic force. On smaller scales, the two forces would appear to behave quite similarly.

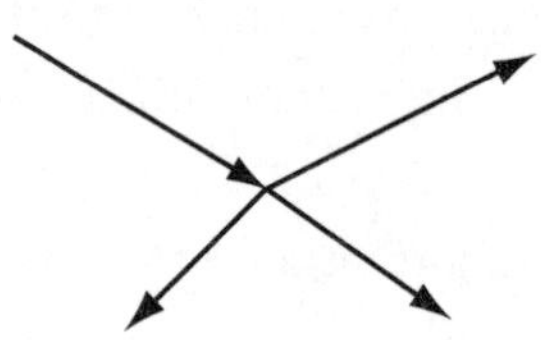

FOUR PARTICLES INTERACTING AT A POINT WITH NONSENSICAL HIGH-ENERGY BEHAVIOR

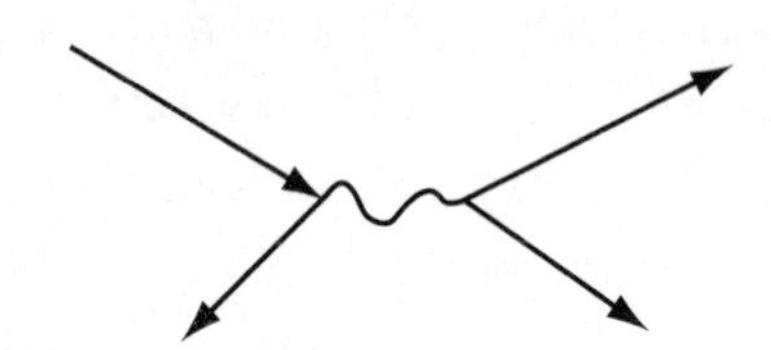

EXCHANGING A VIRTUAL PARTICLE WITH SENSIBLE HIGH-ENERGY BEHAVIOR

String theory had the potential to solve similar nonsensical predictions of the naive quantum version of general relativity. In this theory, recall, the gravitational force occurs because of the exchange of virtual particles, called gravitons. Because of the complicated structure of general relativity, it turns out that there are an infinite tower of possible interactions of gravitons with each other, so that one can find interactions of three, four, five, or more gravitons at a single point.

It turns out that, in a way similar to that in which the interactions of four particles at a single point in the weak interaction produced nonsensical results, these many-particle interactions in general relativity ultimately produce a host of infinities if one allows the energies involved to become arbitrarily large.

But string theory offered a new opportunity to once again change the rules at small distances. If the particle we call the graviton is, at sufficiently small scales, resolved instead to be a vibrating string, then what is allowed at small scales will change. It turns out that, for technical reasons, a graviton is required to be made up of a closed string loop rather than a string segment whose two ends are not connected. In this case, one can redraw what would otherwise appear at large distances to be an interaction involving four gravitons at a single point. The picture becomes more complicated than simply having two graviton particles exchange some other particle with two other gravitons located some distance away, as in the weak case. Rather, one imagines a more complex process in which the vibrating string loops that masquerade as gravitons at large distances bifurcate and exchange other vibrating loops as shown in the second diagram below, which looks like two pairs of trousers sewed together. But while this is more complicated to draw, the effect is the same: The seemingly point-like interaction of gravitons is instead spread out over some region of space, providing a new lower-scale cutoff that yields results that are finite for such physical processes, even as the energies of the particles involved become very large.

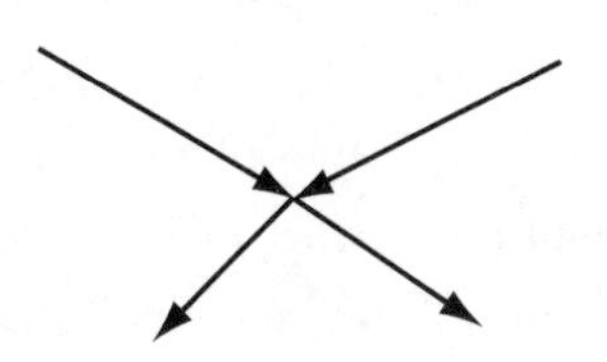

FOUR GRAVITONS INTERACTING AT A POINT

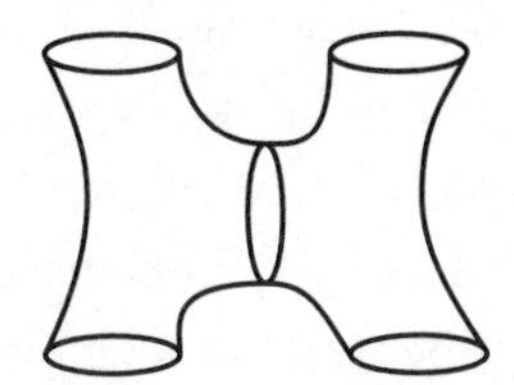

FOUR GRAVITONS AS CLOSED STRING LOOPS INTERACTING OVER SOME REGION BY THE EXCHANGE OF STRING LOOPS

In superstring language there is another way of viewing this effect, and that is that the string has a fundamental symmetry, called "conformal invariance." This symmetry would imply that the physical nature of string interactions is independent of how one might stretch the string. Thus, for example, two strings that might seem to be otherwise close together during an interaction can in fact be stretched farther apart, and one would still get the same answer for the contribution of this process to physically measurable quantities. But, as we have seen, if the interaction points are spread out in space, then the dangerous infinities tend to be removed.

This conformal, or stretching symmetry turns out to have unexpected implications when strings interact in certain exotic spaces. For example, in the particular case where one has closed string loops moving on a space that looks like a donut, called a torus, then it turns out that a string loop having a very small size around one circle of the donut behaves identically to a large loop stretched around the other circle (the circle spanning the horizontal direction around the donut).

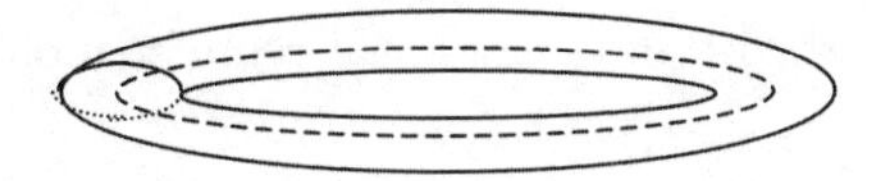

A SMALL LOOP STRETCHED AROUND ONE CIRCLE OF A TORUS AND A LARGE LOOP STRETCHED AROUND THE OTHER

This was a remarkable result and its implication was very important in the attempt to understand why strings might universally tame quantum infinities. For if there is a symmetry that says that string loops of radius smaller than some quantity–say, R_0–produce identical physical effects to those of strings of a size much bigger than R_0, the implication is that R_0 represents some fundamental physical scale below which distances have no physical meaning. If you do try to probe smaller scales using strings that appear to be smaller in size, you end up producing phenomena that could instead by equally well pictured as involving strings of a much larger size. This "duality" between large and small strings, as it is called, can therefore be seen as providing a clear physical cutoff on how small a region can be over which virtual processes can occur. Once again, this small-scale cutoff has the effect of rendering otherwise potentially infinite virtual processes finite.

While spreading out the interactions of gravitons is one way to turn gravity from a quantum theory beset with infinities to a quantum theory that is apparently finite, having *a* finite theory does not imply that one has *the* finite theory. A host of other issues, both physical and mathematical, must be addressed before we might gain confidence that this is the case.

This brings us to the truly unexpected string miracle. It was also discovered in 1984 that the quantum theory of supersymmetric strings in ten dimensions can, in certain circumstances, naturally avoid another type of more subtle and dangerous mathematical inconsistency I mentioned earlier, which physicists call an "anomaly." An anomaly occurs when quantum mechanical virtual processes destroy the mathematical symmetries that one would otherwise expect a theory to possess. It is as if one produced a theory that predicted the earth should be a perfect sphere without any imperfections, so that any place on the planet would be identical to any other place, but when one considered quantum mechanical effects one would instead find that on small scales the sphere would contain mountains and valleys, so that some of its points would be very different than other points. Thus, the beautiful spherical symmetry of the theory would be destroyed.

Such nasty quantum mechanical anomalies have been found to generically occur in one particular type of quantum theory: that which distinguishes left from right. Unfortunately, as we have seen, the weak interaction is precisely such a theory, in which "left-handed" electrons behave differently than "right-handed" electrons.

To step back a bit, it was somewhat of a surprise that strings in higher dimensions even allow for such a possibility of "handedness" in the first place. Careful studies of Kaluza-Klein theories in higher dimensions by Ed Witten, in particular, had earlier demonstrated "no-go" theorems implying that there was no straightforward way to distinguish left- and right-handed fermions in higher-dimensional theories.

It turned out, however, that one can avoid these no-go theorems if one changes the rules a bit. Namely, if instead of pure Kaluza-Klein gravity in the higher dimensions, one supplements the theory by having extra Yang-Mills fields living in these higher dimensions–precisely the situation that, I remind you, arises in supersymmetric string theories in ten dimensions–then these fields can impact upon the fermions living on strings in complicated new ways in order to produce right-handed and left-handed objects that behave differently.

But with this realization came the concern about anomalies. In general, once left- and right-handed fermions behave differently, then the quantum mechanical contributions of virtual left- and right-handed particles to various processes can destroy the very symmetries that are required in order to keep the theory mathematically consistent. These anomalies essentially undo the very careful cancellations of various otherwise infinite quantities that are ensured by the Yang-Mills symmetries, as well as resulting in a host of other nonsensical predictions.

Actually, things are even worse in ten dimensions than in four, because not only can the Yang-Mills symmetries get destroyed by anomalies, so can the symmetries that underlie general relativity. Thus, there is actually a greater chance that any given theory of gravity will prove to be nonsensical as a quantum theory in ten dimensions than it will in four dimensions.

What Green and Schwarz showed in 1984 was that for two specific kinds of supersymmetric string theories in ten dimensions, the theory was not only finite, but even with left- and right-handed fermions acting differently, all anomalies disappear. What might result therefore could be a completely finite and consistent quantum theory of gravity.

Within months of the Green and Schwarz discovery, feverish activity by two different groups produced two more dramatic developments that ultimately generated enough excitement to induce much of the rest of the particle physics community to drop what they were working on and begin to explore this new possible Theory of Everything.

The first development involved a group led by David Gross, who, you may recall, helped to kill the first incarnation of dual string models when he discovered the phenomenon of asymptotic freedom, which demonstrated that QCD, and not a dual string model, was the proper theory of the strong interaction. As I indicated earlier, David's graduate career had begun at Berkeley, and continued at Princeton, where he did important work on dual string models with Neveu, Scherk, and Schwarz. His return to this subject, after having abandoned it a decade earlier, was nothing short of triumphant, and he has taken it up again with all the fervor, as he himself suggested, of a converted atheist.

Gross, along with his colleague Jeff Harvey and students Emil Martinec and Ryan Rohm, developed, in a tour de force, something with the memorable name of "heterotic string." The name does not derive from the word *erotic,* but rather from the root *heterosis,* although there is also no doubt that the model is kinky, both metaphorically and literally. Indeed, it is so imaginative as to be considered sexy by many theorists, which says something either about the model or about theorists.

When Green and Schwarz discovered that superstring theories in ten dimensions could be consistent, finite, and anomaly-free, they identified two possible symmetries of strings that would allow this. They explicitly demonstrated three different sorts of superstring solutions that exhibited one type of symmetry, but none that exhibited the other type, which for a number of technical reasons seemed like it might produce more interesting grand unified scenarios. The heterotic string, on the other hand, could work with either symmetry and thus was of special interest.

What made this particular string theory so exciting, however, was not merely that it could produce potentially more interesting Yang-Mills symmetries, but that the existence of this Yang-Mills symmetry was forced upon it, not by the seemingly ad hoc need for anomaly cancellation, but by the requirements of formulating the string theory itself. This suggested some potentially deep connection between the possible existence of strings in ten dimensions and the observed Yang-Mills symmetries of nature in four dimensions.

The heterotic string model involves closed string loops, which on first glance is unusual, because closed strings, while they incorporate gravity, do not generally incorporate Yang-Mills symmetries. Gross and his collaborators, however, realized that if one is bold enough then this limitation can be circumvented. In particular, on a closed string, the vibrations that travel in one direction around the string are completely decoupled–that is, they do not interact with the vibrations that travel in the other direction around the string. There is a classical analogy for this: If you take a regular string, and jiggle it from one end to send a wave down it, while at the same time jiggling it from the other end to send a wave in the opposite direction, you will be able to see the two waves pass directly through each other at the center of the string. The two wave modes do not interact.

Now for an amazing feat of mathematical sleight-of-hand: If it is possible to imagine a sort of "hybrid" string in which the left-moving and right-moving vibrations on a string are quite different. In fact, Gross and his coworkers argued that these different modes could actually be pictured as living in different sets of dimensions!

For the ten right-moving sets of vibrations on strings, Gross and colleagues treated them precisely as Green and Schwarz did for their ten-dimensional superstring: with ten normal coordinates, and with sixteen of those strange Grassmann anticommuting coordinates. Recall that the effect of this construction is to produce equal numbers of fermion and boson excitations on the string.

Gross and coworkers then imagined that the poor left-moving vibrational excitations were bereft of supersymmetry. You may recall, however, that the quantum mechanics of vibrating strings without supersymmetry can only be formulated consistently in twenty-six dimensions. In a leap of creative chutzpa that is hard to beat, Gross and his colleagues then simply imagined that the left-moving vibrations on strings act as if they live in a twenty-six-dimensional space!

It may seem strange to you that some of the vibrational modes on a closed string live in one number of dimensions, while others live in another, much larger set of dimensions. Actually, this little technicality was not lost on the creators of the model, who pointed out an apparently straightforward, if equally bold, solution. Simply curl up sixteen of the dimensions on which the left-moving vibrations operate into very small regions. In this case, then, just as happened in Kaluza-Klein theory to make the fifth dimension invisible, on scales too large to resolve the extra sixteen dimensions, one would appear to be left with only ten remaining left-moving modes to go along with the ten right-moving modes.

In a way this mathematical wizardry is also reminiscent of what happened in the original Kaluza-Klein model. There, degrees of freedom in the extra curled-up dimension end up looking, in the four-dimensional world, like photons (i.e., particles associated with the gauge symmetry of electromagnetism). In the new model, one could show that the extra sixteen left-moving modes associated with the curled-up sixteen dimensions end up appearing as welcome extra Yang-Mills symmetries and fields on the remaining ten-dimensional closed string.

Incidentally, if this isn't strange enough for you, it turns out that there is a way to frame the heterotic string in which the extra sixteen left-moving modes are not associated with sixteen extra spatial dimensions at all, but rather with thirty-two extra weird Grassmann anticommuting coordinates on a ten-dimensional string! In string theory, it seems, as we shall see again later, the existence of extra hidden dimensions may actually depend upon the eye of the beholder.

This brings us to the final development of the trilogy in 1984–85 that truly put string theory at the center of the particle physics universe. With the excitement generated by finite, consistent superstrings and the new heterotic possibility of generating large and phenomenologically interesting Yang-Mills symmetries in ten dimensions, there was only one tiny thing left to do: Make some contact with the four-dimensional universe of our experience!

Enter Ed Witten. While some time earlier, Claude Lovelace at Rutgers had begun to examine what might happen if one put strings on spaces that curve up into small balls, a comprehensive analysis of how one might turn these hypothetical hyperdimensional theories of everything into realistic models of our world was carried out first by Witten, and then in a seminal paper by Witten and his collaborators Philip Candelas, Gary Horowitz, and Andy Strominger.

Witten first showed that one could in principle "compactify" six of the ten dimensions associated with superstrings into small, finite volumes in a way that would leave four large dimensions left over while still preserving, in those four dimensions, essential features such as the absence of anomalies. Then, Witten and his colleagues, a second "string quartet," (the first being Gross and colleagues) explicitly demonstrated how this might be done. The key was to rely on a new type of mathematics, not then well known among physicists, called "Calabi-Yau manifolds," after the mathematicians who had first described them.

A "manifold," in mathematics, is something like a rubber sheet. Generalizing the properties of such smooth, pliable objects to higher dimensions has allowed mathematicians to invent a host of strange new objects. Calabi-Yau manifolds are one interesting mathematical class of manifolds with exotic curvatures in many dimensions that can be mathematically classified.

Remember that Kaluza and Klein had considered the simple case where their single extra dimension was curled up into a small circle (a very simple one-dimensional manifold). One might likewise imagine that this concept could be applied to the six extra dimensions in string theory, have them curl up into a small six-dimensional sphere. This was the approach first explored by Lovelace, but it turns out not to work. As Witten and collaborators demonstrated, very specific conditions that needed to be imposed on this "compactified space" in order for the resulting four-dimensional theory to remain sensible.

Such spaces turned out to have been investigated by the mathematicians Eugenio Calabi and Shing-Tung Yau, and Witten (who would later win the most prestigious award in mathematics, the Fields Medal, for his work using string theory to illuminate the detailed mathematics of knot theory) and his collaborators were able to use their results to explore what kind of theories one might expect to produce in four dimensions.

The results were encouraging. It appeared to be possible to produce theories with plausible grand unified Yang-Mills symmetries, and with a spectrum of elementary particles, quarks, electrons, muons, and so forth that could bear an eerie resemblance to what we actually observe in our universe.

The reaction to the Candelas and coworkers paper by the physics community was astounding. Suddenly the esoteric and mathematically complex field of string theory held the promise of actually making contact with reality–and not just slight contact. It opened up the possibility of providing a fundamental explanation of why the world at its most basic scale looks like it does, and the answer seemed to lie hidden in this extra-dimensional Calabi-Yau universe.

Within two to three years most major physics departments had a group of brilliant young theorists working on string theory, and in turn this group, usually tenured within a few years of getting their PhDs, started training a new generation, many of whom began their training with string theory, and had never heard of such elementary particles as pions, which had started the whole effort off in the first place.

It was a common belief at the time that even though the theory was so complex that the approximations that had thus far been performed barely scratched its surface, it was just a matter of time–and not much time, perhaps–before all the details would be worked out and all the big questions answered. For example, in order to approximate the complex Calabi-Yau manifolds, physicists instead explored approximations called "orbifolds," which on the whole behave like higher-dimensional generalizations of the nice, smooth rubber sheets one can picture in one's head, but which have, at a discrete number of points, locations where the sheet gets warped into a conelike shape, with a single point of very high (in a strict mathematical sense, infinite) curvature. Thus, all of the complexities of the Calabi-Yau manifolds could be relegated to what might occur at a finite number of weird points in an otherwise smooth and simple space. One hoped that big questions would be insensitive to this dramatic approximation.

In the meantime, the world of elementary particle physics underwent a sea change after 1984. In particular, an interesting sociological phenomenon began to take place that still has repercussions for the field today. The largely mathematical questions underlying the new theories became for a number of young physicists new to the field much more interesting than trying to figure out such "trivial" low-energy details as how grand unification might account for the actual physics that resulted in a universe full of matter instead of antimatter, or why the proton is two thousand times heavier than the electron. In short, the as-of-yet hypothetical world of hidden extra dimensions had, for many who called themselves physicists, ultimately become more compelling than the world of our experience.

CHAPTER 15
M IS FOR *MOTHER*

I never think of the future. It comes soon enough.

–Albert Einstein

The theoretical discoveries of 1984–85 energized theoretical particle physicists as nothing had done in a long while. At the same time they produced a remarkable optimism in those who had already begun to work on string theory that the long-sought goal of a consistent unified theory of all the fundamental interactions in nature was at hand, if only the theory could be fully understood. What began as an investigation of an idea that might incorporate gravity and quantum mechanics had, precisely because of its enforced necessity of extra dimensions, begun to appear as if it might explain why everything else existed as well.

The concluding sentence of the original heterotic superstring paper stated, "Although much work remains to be done, there seem to be no insuperable obstacles *to deriving all known physics* from the . . . heterotic superstring." (Italics mine.)

With the realization that the heterotic superstring literally required, for its internal consistency, precisely those Yang-Mills symmetries that appeared most promising to describe the real world, it seemed as if nature was saying, "Build a string, and they will come." If the requirements for a consistent string theory in turn required a specific Yang-Mills symmetry that might explain all of the observed distribution of particles and forces in our four-dimensional universe, then maybe we could finally resolve Einstein's long-ago query, "Did God have any choice in the creation of the Universe?" The answer would be "No, not if she chose to create it via strings!"

Along with the optimism came a sense of astonishment: Within the course of less than a year it seemed as if an almost insurmountable problem had largely been resolved. So it was that one often heard the remark that, by means of a fortunate accident (the development of dual string models to attempt to explain the strong interactions) we had discovered what rightfully should have been considered twenty-first- or twenty-second-century physics in the twentieth century. We were truly living in the future!

And if life were an impressionist painting, we would have been. Seen with broad brush strokes, everything appeared to be in order. However, there were still a number of nagging details, not to mention the growing recognition that the theory was nowhere near to being fully explored, let alone understood. Indeed, it was not quite clear precisely what string theory actually *was.* In a prescient paper written in 1983, shortly before the great string revolution, in which he guessed that string theories might be candidates for a consistent theory of quantum gravity, Ed Witten admitted, "What is really unsatisfactory about string theory at the moment is that it isn't yet a theory."

Unfortunately, the closer one looked, the greater the problems became. The very richness of the string models and compactification schemes, for example, appeared to undermine claims for uniqueness and with it the hope that string theory would prescribe a universe that simply *had* to look precisely like the one we live in. Shin-Tung Yau had, for example, elucidated over a hundred thousand different Calabi-Yau manifolds, and compactifying six dimensions on each of them would produce a different four-dimensional theory.

Moreover, detailed analysis of the approximations used to compactify the theory from ten dimensions to four suggested that these operations might not be well controlled, invalidating the attractive phenomenological pictures that had first been presented. In an effort to check whether the four-dimensional theories that appeared to result from compactification really were consistent, theorists began to analyze string theories in four dimensions from a new perspective.

It turns out that because a string is a one-dimensional object moving in time, its "world sheet"–that is, the region of space-time it maps out as it moves–is a two-dimensional surface. This is the case whether the string is moving in four dimensions, ten dimensions, or twenty-six dimensions. Adding new fields onto the world sheet, which is what happens when fermions and Yang-Mills fields are added to strings, therefore involves studying how fields behave on two-dimensional surfaces.

Interestingly, this is an area of intense interest in condensed matter physics, which studies the bulk properties of real material, whether boiling water, superconductors, or magnets. When such materials undergo a change of phase–for example, water begins to boil, magnets become magnetized–then near the point of this change the properties of the material become particularly interesting and simple. The physics turns out in some cases to depend almost entirely on phenomena associated with two-dimensional surfaces, such as bubble walls form the boundary between different phases of boiling water. As a result, condensed matter physicists have become experts on studying such surfaces. Moreover, it turns out that as materials approach the conditions where such phase transitions can occur, their nature begins to look self-similar (i.e., the same phenomena like bubbles seem to appear on all scales). This "scale invariance" is similar to the conformal symmetry of the string theories, which implied that the physics looked the same regardless of over what scales one might stretch the strings.

In any case, studies of such condensed matter systems had classified essentially all two-dimensional field theories, and demonstrated that many of them had the properties that one guessed they might have if they instead described string world sheets obtained by compactifying from higher dimensional theories. That was the good news. But at the same time it suggested that perhaps one could consider string theories in four dimensions without ever worrying about their ten-dimensional roots. Indeed, are the ten dimensions necessary at all, or are the extra dimensions just mathematical artifacts? This is the central question that continues to haunt us.

It was clear that to go beyond the impressionistic connection to the real world, one was going to have to understand string theory a lot better than it was understood thus far. And this was going to be hard work, involving the development of new mathematics that could handle systems far more complex than anything that had been heretofore studied.

An army of bright new physicists immediately launched a campaign to scour every cave where interesting possibilities might lurk. Over the next four years the line that had previously tended to separate articles that appeared in physics journals from those that were published in mathematics journals began to blur. Ed Witten, in particular, worked furiously on a host of remarkable ideas.

But, in spite of this plethora of talent and output, progress in actually answering questions about our four-dimensional world was distinctly lacking. New insights about the possible nature of string theory, field theory, and Yang-Mills theories might have been accumulating, but solid physical predictions were not.

Most embarrassing (from my point of view, at least) was the apparent inability of string theory to address the key physical paradoxes that seemed to be associated with a quantum theory of gravity. Sure, the theory appeared to get rid of infinities that might otherwise render predictions nonsensical, but when it came to predicting such things as what the energy of empty space might be (i.e., why the cosmological constant must be zero or extremely small), the theory appeared to make no useful predictions.

Another area where strings had thus far shed no light was the very question that Stephen Hawking raised that appeared to result in a direct challenge to quantum theory itself in a world of gravity. What happens to the information about what falls into black holes if the black holes can ultimately evaporate away and disappear?

While not much had happened on these fronts, theoretical progress in trying to understand the different varieties of consistent string theories had begun to suggest that the five different types of consistent string theories explored in ten dimensions, might be related.

Might these apparently different theories merely be different manifestations of some single "*über*" theory? As early as 1985, in fact, several researchers had suggested this possibility. After all, this is precisely the trend that had worked so well to simplify the physics of the known world: Electricity and magnetism had been shown to be different reflections of the same force, the weak and electromagnetic interactions had been shown to be different reflections of the same underlying physics, and so on.

Interestingly, however, when physicists began to explore such a possible new connection between the different string theories, hints began to appear that these different theories might well be unified–but not in ten dimensions. Rather, they seemed as if they might be different ten-dimensional reflections of an underlying eleven-dimensional theory!

Alert readers may remember that eleven dimensions had previously appeared in the grab bag of theoretical physics, associated with a special theory of supergravity. In eleven dimensions, all interactions and particles are specified by gravity and supersymmetry alone, while in ten dimensions there is much more freedom to choose extra Yang-Mills symmetries, fields, and so on. Perhaps an eleven-dimensional theory might be unique, even if a ten-dimensional theory wasn't.

The first step on this road came from work by Witten and collaborators in 1995s, which suggested that all five known consistent string theories were merely different versions of a single underlying, more expansive theory. The next major development in understanding this possible unification came from a remarkable and unexpected observation in 1995 by Joe Polchinski, at the Kavli Institute for Theoretical Physics at Santa Barbara.

Polchinski changed the whole nature of our understanding of what was possible in string theory because he demonstrated that what people had been exploring up to that point–indeed, the theory that had been claimed to be a theory of *everything*–had in fact overlooked an infinite number of things, including new objects in higher dimensions. For reasons that will become clear, he called them *D-branes*.

His observation derived from considerations of how open strings might behave in toroidal (i.e., donut-shaped) spaces. As you will recall, in such spaces it appeared that shrinking one radius of the donut produced a theory that, for closed strings that might wrap in different directions around the donut, looked identical to one in which the same closed strings were wrapping around a radius that became very large.

Open strings–that is, strings that do not close back upon themselves, forming loops, but have two end points like a regular piece of string–however, end up in this case leading to another interesting phenomenon. Their ends are free to move about, and it turns out that the surfaces comprising the set of points along which their ends can move can themselves form a whole new type of mathematical object, behaving like a sort of (mem)brane. In three spatial dimensions, for example, a two-dimensional brane could be a plane or a membrane surface like a rubber sheet. Open strings would be attached at either end to this plane (as the diagram shows). They could wiggle and move in the extra dimension, but their ends would by definition, move about on the plane (brane).

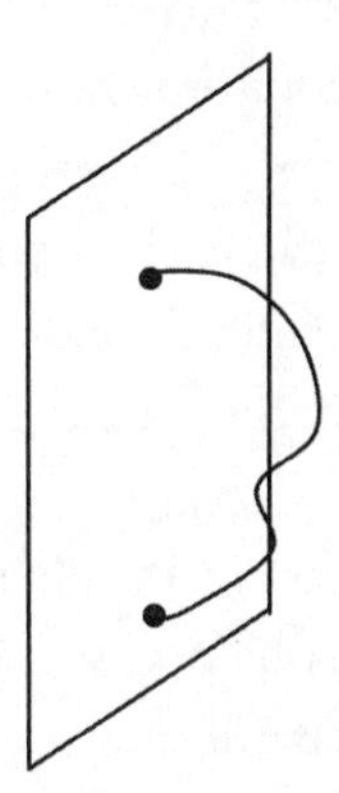

One might imagine that these structures are called D-branes because they need not be two-dimensional, but can be any number of dimensions, less than or equal to the total number of dimensions of space-time itself. That would be too simple, however. It turns out that they are called D-branes because of the special mathematical conditions (called "Dirichlet boundary conditions") that one imposes, which Polchinski realized could exist when a string ends on a surface. The different dimensional D-branes are actually called "p-branes" (since the letter *D* was taken already), where *p* refers to the dimensionality and *D* to Dirichlet. A one-brane looks like a string, a two-brane looks like a familiar membrane (like a rubber sheet), a three-brane like our own three-dimensional space, and so on.

What is more notable about these new objects than their names is the fact that they have their own type of dynamics. Recall that years earlier, when dual strings were first being explored, physicists had wondered whether one might generalize the underlying concepts to yet higher-dimensional objects. In a sense, Polchinski's D-branes are just these generalizations, but more interestingly, he showed that they are *required* to arise when one attempts to consider the full dynamics of string theories.

They had been previously missed for two reasons. First, almost all of the previous analyses of strings had dealt with the simplest approximation to the theory, the so-called weak coupling limit–namely, when strings are almost noninteracting and their wiggles are minimal. Second, fixing the ends of strings to lie on some surface spoils some of the space-time symmetries of the theory in ten dimensions. Physicists had tacitly assumed that keeping such symmetries was essential. But they seemed to forget that the world we experience is only four-dimensional, and what is important is that the resulting theory have the observed space-time symmetries in four dimensions that Einstein ultimately incorporated into general relativity. D-branes, through the mathematical conditions that occur when strings are connected to them, preserve these latter symmetries, if not the full ten-dimensional symmetries.

Once D-branes are included in the theory, it becomes much richer and more complex than it was before, with a host of possible new phenomena. One might imagine that it was somewhat of an embarrassment that string theorists had previously proclaimed that they were on the verge of victory in creating a "theory of everything," when they had in fact virtually missed "almost everything" in the theory. But in the ever-optimistic string worldview, there are no embarrassments.

On a slightly less facetious note, it is important to realize that devoting literally decades of one's career to a theoretical struggle, with unknown odds for success, requires those who engage in it to have a deep underlying faith in the validity of what they are attempting. For these "true believers," every new development provides an opportunity to confirm one's expectations that these ideas ultimately reflect reality. What separates this from religion, or what *should* separate this from religion, however, is the willingness to give up these expectations if it turns out that the theory makes predictions that disagree with observations, or if it turns out that the theory is impotent and makes no predictions.

In any case, what made D-branes a cause for celebration rather than sullenness, was that they allowed a full demonstration that the various consistent string theories in ten dimensions were in fact different aspects of the same theory. In order to establish this, the previously discovered "duality" of open strings on donut-shaped toroidal spaces–in which large and small radii of the different compactified dimensions are exchanged–was essential. Once D-branes are included in the picture, going to the small-radii limit in one type of string theory could be seen as producing the same physics as the large-radii limit of another theory.

D-branes are also of great interest because charges can exist on them, like electric charges, that are the source of fields like the electromagnetic field. Since D-branes are the surfaces on which the two ends of open strings are fixed, and it turns out that Yang-Mills charges can exist on the ends of open strings, these charges are then fixed to the D-dimensional surface of the brane. However, remember that closed strings, which have no end points and thus are not fixed to branes, also incorporate all the physics associated with gravitons, the particles that convey gravitational forces. Thus gravity can operate throughout the "bulk" ten-dimensional space both on and between the branes, while the charges that are the source of Yang-Mills fields live on the branes themselves. As we shall see, this can have dramatic implications.

In any case, the presence of D-branes in string theory also allowed theorists to explore the all-important domain where strings might interact strongly with one another, an area that could not be addressed using conventional techniques developed to try to understand the theory. This was especially critical because it was known that considering only the possibilities where strings might interact more feebly with one another would yield a picture of the theory that was not fully accurate, quantitatively or qualitatively.

In particular, it was discovered that there is a new kind of "duality" in string theories with D-branes. Recall once more that for strings living on toroidal (i.e., donut-shaped spaces) the large-radius physics is equivalent, and thus "dual" to the small-radius physics. Now, when D-branes are introduced into the picture, a new and different sort of duality results that connects what otherwise may seem to be disparate physical extremes, obtained by interchanging strings and branes in the theory. This interchange maps a part of the theories where strings may be interacting strongly with each other, and where one cannot perform calculations, with a part of the theories where the strings are more weakly interacting, and their behavior can be more simply followed. In this way, not only might one hope to explore new features of the various different string theories, but it becomes possible to demonstrate how different theories might be related.

The good news is that a new relation between formerly disparate theories was uncovered. The bad news is that while previously there had been five distinct consistent string theories–suggesting that string theory in ten dimensions, with six dimensions ultimately being compactified to leave four large dimensions, was not unique–there now appeared to be a continuum of theories. Specifically, these different theories were related to one another, but each theory represented a distinctly different physical limit. These different theories could be continuously transformed into each other, implying a continuously infinite number of intermediate physical possibilities.

There was a ray of hope, however. When examining one of the string theories with branes when the string interaction strength became large, the number of states grew in such a way that it appeared as if some new, hidden dimension was appearing. Recall that in the original Kaluza-Klein theory, as long as one was considering distances much larger than the radius of the circular fifth dimension, all the extra five-dimensional degrees of freedom remained hidden. However, as the radius of the fifth dimension becomes larger in this model, the energy required to resolve these new states decreases. Ultimately, as the radius goes to infinity, the infinite tower of new states makes its presence known. Such behavior was precisely what was being observed for the number of D-branes in this string theory as one tuned up the string interaction strength. Suddenly an eleventh dimension began to suggest itself.

This apparent extra dimension was not observed in the first decade following the superstring revolution in 1984 precisely because the analysis of weakly interacting strings could only reveal a small part of the theory. It was now understood that this "weak-coupling" approximation was really very similar to what our four-dimensional world is in the original Kaluza-Klein model–namely, an approximation to reality obtained when the size of the extra dimension is very small compared to anything one might measure. It would have been missed, just as a fifth dimension would be forever missed in the original Kaluza-Klein model, if one always did experiments on scales much larger than the extra compact dimension.

This is as close as anything can come to "physics irony." Here we had an apparently remarkable new paradigm for physical theory that in some sense had ultimately been motivated by the suggestion of Kaluza and Klein that the physics of our world might derive from the hidden physics of extra dimensions. Yet hidden within the theory itself apparently lies hidden physics of yet another hidden dimension!

The key questions then become: What is this new hidden physics, and does the propagation of dimensions continue? The answer to the first question was, and to some extent still is, "Anyone's guess." Clearly the theory will in some limit in eleven dimensions resemble supergravity, which forms the basis of much of string theory. But at higher energies it is unlikely to resemble either supergravity or string theory, but perhaps something even more miraculous.

One thing is clear, however. If this picture is correct, what string theorists had previously claimed were *fundamental* tiny strings wiggling in tiny extra dimensions deep inside what we otherwise thought were fundamental elementary particles, would in fact perhaps be tiny membranes wrapped around yet other tiny extra dimensions, with yet even more fundamental objects. They would be masquerading as strings because, in the approximations that had been used to define the string theories in question, the extra dimension was curled up on a scale smaller than the string scale, so that a two-dimensional surface would look like a one-dimensional string. Strings, in this respect, need not therefore be the truly elementary objects in the theory.

Even when a new theory might not be understood fully, at least it can be labeled. This new eleven-dimensional theory has become known as M-theory. What does the *M* stand for? Well, first we must recognize that the term *M-theory* has evolved to encompass not just the theory that the ten-dimensional theories each approach as some parameters are varied, but the theory that encompasses all the theories in all their limits! Thus, it is only partially facetious to claim that the name stands for "mother of all theories." I am told that Ed Witten introduced the term and said it stands for magical, or mysterious, but that may be apocryphal. Other proposals exist: Membrane theory? Marvelous theory?

A somewhat more informed guess, however, suggests that perhaps the *M* stands for *matrix*. The argument for this is based on the fact that if one takes one of the string theories that appears to suggest this hidden extra dimension, then as the string interaction strength is varied, the quantities that would normally be the coordinates describing the motion of the strings and branes are not simple numbers but are instead described by mathematical objects called matrices.

A matrix is like a table of numbers, arrayed in rows and columns. Here are two examples:

$$\begin{pmatrix} 1 & 5 & 7 \\ 3 & 4 & 8 \\ 2 & 3 & 6 \end{pmatrix} \quad \text{AND} \quad \begin{pmatrix} 2 & 4 & 5 \\ 1 & 6 & 3 \\ 4 & 2 & 9 \end{pmatrix}$$

Matrices can be treated like ordinary numbers in that one can define for them operations such as multiplication and addition. However, unlike normal numbers, matrix multiplication is not commutative. That is, while 3×4 equals 12 whether or not one multiplies 3 times 4 or 4 times 3, the product of two matrices A and B is *not* in general equal to the product of B times A. This is because the rules for multiplying matrices are complicated. One multiplies each term in the first row of one matrix times the term in the corresponding column and then adds the sum to get the corresponding term (upper-left-hand corner) in the new matrix. Thus, for example, for the two matrices given above, the first term in the corresponding matrix if I multiply the first matrix times the second is $[(1 \times 2) + (5 \times 1) + (7 \times 4)] = 35$. However, if I multiply the second matrix times the first, the first term in the new matrix is $[(2\times1) + (4\times3) + (5\times2)] = 24$.

What is interesting and at the same time odd about this is that if matrices are the fundamental objects describing the eleven-dimensional universe of M-theory, then each point in the space is described by a matrix and not a mere number. This means the eleven-dimensional universe of M-theory bears little or no resemblance to the universe we experience. The coordinates that describe where you are in this space don't commute with each other! As if eleven-dimensional ordinary space was not complicated enough to think about.

Equally important is the fact that in this new eleven-dimensional space, neither strings nor D-branes may be the truly fundamental objects. If this picture is correct, strings in ten dimensions are just as much an approximate illusion of reality as elementary particles in four dimensions were supposed to be in the original string picture.

One might, of course, wonder if all of this rampant breeding of new dimensions is any different from the earlier rampant breeding of new elementary particles at ever higher energies, which seemed so confusing and complex in the 1960s, and which led, in a sense, to the original proposal for dual string theories.

Nevertheless, there are reasons to suspect that eleven dimensions are as far as one need go. After all, one cannot have sensible supergravity symmetries in higher dimensions, and supergravity is one of the hallmarks that is supposed to characterize feasible and consistent string theories as candidates for quantum gravitational theories.

Readers with a fantastic memory and remarkable attention for detail may remember that another feature of eleven-dimensional supergravity theories was that gravity determined all of the matter fields in the theory, and that there was no room left over for Yang-Mills fields and all the other paraphernalia that makes our world so interesting. So, what is the difference in M-theory? It is that M-theory contains many more objects than merely elementary particles and fields. It contains things that look, in some limits, like strings and D-branes, and in other limits, like matrices. And who knows what else?

Finally, after this seemingly miraculous convergence on an unknown

M-theory (I remind you that for some people everything in string theory is miraculous), you might think that this fiddling with extra dimensions would be over with. However, the next, and up to the present time, last string miracle was yet to occur.

In 1997, a young Princeton graduate student turned Harvard professor Juan Maldacena made a daring conjecture, which once again completely changed the face of string theory. Remember that strings in ten dimensions can host Yang-Mills gauge fields, while in eleven dimensions at low enough energies, gravitational degrees of freedom associated with supergravity are all that can be detected. Maldacena suggested another kind of dramatic correspondence appropriate for our understanding of Yang-Mills theories in four dimensions (i.e., the world of our experience). Using ideas based in ten-dimensional string theory, Maldacena proposed that perhaps our four-dimensional world, full of Yang-Mills gauge symmetries, might have a hidden five-dimensional meaning. Specifically, he conjectured that a four-dimensional flat space with quantum Yang-Mills fields and supersymmetry, which our world might contain, could be completely equivalent to a somewhat strange five-dimensional universe with just classical (super)gravity and nothing else.

If this sounds suspiciously like déjà-vu all over again–namely, like a modern reframing of the original Kaluza proposal of 1919, in which electromagnetism in four dimensions arose from an underlying theory involving just gravity in five dimensions–you are not that far off. But there is a fundamental and critical difference. In Kaluza-Klein theory, and all subsequent theories with extra dimensions, our four-dimensional universe is merely the tip of the iceberg. We only see four dimensions because our microscopes cannot resolve those tiny extra dimensions. However, in Maldecena's conjecture, four-dimensional space is not just some large-distance approximation of the underlying five-dimensional space. Rather, the two are precisely the same! All the physical laws of one universe are equivalent to those of the other universe!

Before wondering what this idea might imply regarding the actual meaning of extra dimensions, you might wonder how it could be possible that four dimensions could contain *all* the physical information of a five-dimensional universe? After all, if one has extra dimensions, there are extra physical degrees of freedom available. In our own world, for example, it is hard to ignore the extra freedom offered by being able to access the third dimension to jump over obstacles on the ground, or the second dimension to go around obstacles in front you.

If four dimensions are somehow to encompass five, then somehow the extra five-dimensional physical degrees of freedom have to be encoded—obviously in a different form—in the lower-dimensional space. Perhaps it is simplest to think of the four-dimensional universe as the surface of a five-dimensional volume. Then the question becomes: How could one encode all the information associated with some volume on a surface bounding that volume?

Framed in these terms, there is a well-known example of precisely this phenomenon in three dimensions: holograms. A hologram, stored on a piece of film or plate, is a two-dimensional record of a three-dimensional scene. But when you look at or through the holographic sheet, depending upon its type and the source of light, you see the entire original three-dimensional image. If you move your head, you can look *around* foreground objects to see objects in the background. Unlike a photograph, which simply stores a two-dimensional projection of the three-dimensional image a hologram stores *all* the information in an image.

The reason a hologram allows this degree of image reconstruction is reminiscent of the information loss problem when material falls into a black hole. If the black hole evaporates, then all the energy that fell into it may be radiated away by Hawking radiation, but the question of whether the information can be retrieved comes down to delicate issues having to do measurement, and what can be reconstructed from subsequent detection of this radiation.

When an ordinary camera records an image, it simply records the intensity of light of each color impinging on the photographic film, or the electronic digital recording media, in the case of digital cameras. Because light is a wave, however, not only does it have an intensity, but its electromagnetic fields at any point oscillate in time as the wave passes by. Different light rays, associated with different electromagnetic waves, will cause electromagnetic field oscillations which will in general be out of phase with one another when they pass different points. This phase information is not recorded when the light intensity alone is measured at any given point. However, holograms manage to use sophisticated techniques to capture this additional information. When this phase data is stored on a two-dimensional piece of film, it turns out that a full three-dimensional image can be reconstructed.

The idea that is central to the Maldacena conjecture—that somehow all the physical information in a volume can be encoded on its surface—has thus become known at the *holographic principle.* I stress that while it has been applied in a variety of contexts by various theorists, the actual Maldacena conjecture itself involved two very specific spaces: a four-dimensional flat space with supersymmetry and quantum Yang-Mills fields, and a five-dimensional space with classical supergravity, along with a very weird specific source of gravity throughout the five dimensions (empty space full of negative energy—unlike anything we have measured in our own universe). Such a space is called an Anti-de Sitter space.

In any case, if Maldacena's conjecture is correct—namely, that there is absolutely no physical difference other than appearance between these two spaces—then the physical distinction between different dimensions itself gets blurred. A host of questions naturally seems to arise. What is the utility of an extra hidden dimension if ultimately nothing is hidden except the existence of the extra dimension? And what is the practical meaning of extra dimensions if you can experience all there is to experience without actually moving into them?

Moreover, we may find ourselves somewhat like the holodeck characters in *Star Trek,* who have no sense that they may be mere projections. Are we merely a pale reflection of the real world behind the mirror? Or, if the surface contains everything that is inside, is it the extra dimension itself that is illusory? If the world of our experience is a hologram, where does the illusion end and reality begin?

Ultimately, if Maldacena's conjecture is correct, then it implies that these questions, as fascinating or troubling as they may seem, are moot. Reality is in the eye of the beholder. Both worlds are real, and identical, as different as they may seem.

If your head is now spinning, it should be. In one chapter, you have been treated, or perhaps subjected, to a menagerie of mathematical marvels associated with strings and D-branes in ten dimensions, M-theory in eleven dimensions, and holography in five dimensions. New dimensions have magically appeared and disappeared with more aplomb than the Cheshire Cat and with an uncanniness that might appear to make Alice's voyage in Wonderland pale in comparison. Most importantly, you may be wondering what all of this wizardry has wrought? Are these imaginings of theoretical physicists any more real or of any more utility than those of Lewis Carroll?

These are good and valid questions. Remember what ostensibly caused all of this mathematical effort in the first place. String theory, or rather the Theory Formerly Known as String Theory, must, if it is to be useful to physicists, address some concrete physical problems and make concrete physical predictions. In its original form, it had simply failed to do so, all the hype surrounding it notwithstanding.

So, as mathematically remarkable as M-theory might be, or as useful as the Maldacena conjecture might be for trying to solve difficult mathematical problems associated with Yang-Mills theories, unless all of these ideas eventually help resolve fundamental physical questions, it *is* all just mathematics.

Thankfully, however, there has been some progress. In my mind it is not clear that it fully justifies the periodic hubris associated with string theory, but we shall see. It is at least an encouraging beginning.

You will recall that a central problem in quantum gravity, which early work on string theory did not appear to address, was the "black hole information loss paradox." Do black holes violate quantum mechanics? And if not, where does all the information that falls into black holes go?

A new approach to this problem did become possible once D-branes began to be explored. Recall that D-branes allow a new connection between the strongly interacting phase of some theories and the weakly interacting phase, where reliable calculations might be performed. It turns out that in certain limits one finds objects in string theory that resemble black holes, with highly curved geometries (in the extra dimensions).

These are called black p-branes. Interestingly, if one explores a different limit of the same theory, where strings are weakly interacting, one can describe much of the physics in a calculable way using standard D-branes. One can hope then that the results of calculations one can explicitly perform in the one limit of the theory where such calculations are feasible might also be applicable in the other limit of the theory, where one cannot do direct calculations, and where the strongly gravitating black p-brane description applies.

Now, if one examines a very special sort of five-dimensional p-brane, then in the weakly interacting limit of the theory, where D-brane calculations become reliable, it turns out that one can explicitly count the number of fundamental quantum states that could be occupied by an object that would, in the strong coupling limit of the theory, be associated with a black p-brane.

The result is striking. The number of quantum states turns out to be precisely the number of states needed to encode the information that was supposedly hidden behind the event horizon of a black hole–the so-called Hawking-Bekenstein entropy. This would suggest that the information is not, in fact, lost down the black hole, but is instead somehow preserved and if we had a way of accurately treating the quantum mechanics of realistic black holes (which, I remind you, are not to be confused with the very special five dimensional black p-branes in this idealized calculation), we would uncover it.

Note that this result is far from a proof that black holes in string theory must behave like sensible quantum mechanical objects, nor does it provide any hint of what might actually happen to the information stored in a black hole's interior as it evaporates. Moreover, the black p-branes in question are actually very finely tuned objects, which wouldn't themselves even evaporate by Hawking processes because of their special configuration.

However, this calculation is at least very encouraging. In the regime where D-branes, which are perfectly well-behaved quantum mechanical objects, are the appropriate description of string theory/M-theory, there are precisely the correct number of states to account for what one might hope a well-behaved quantum mechanical accounting of black holes might require. This was a real computational success in string theory, and it has generated tremendous enthusiasm.

Nevertheless, a host of caveats remain. As one increases the strength of the interaction needed to move from the D-brane to black p-brane picture, the physics could change, and information could be lost. Until one can calculate precisely where the information flows in the evaporation process of realistic black holes, extrapolating the apparent success of this aspect of the theory remains a conjecture.

Also, as I have mentioned, a few months before this writing Stephen Hawking made headlines throughout the world by retracting his claim that black hole evaporation destroys information. He has claimed that a new computation he has performed in the context of classical general relativity demonstrates explicitly how the information that falls into black holes gets preserved as they evaporate. He has spoken about this at several meetings. Many physicists are skeptical. However, when it comes to black holes, Stephen has a good track record.

If Hawking's new claim is correct, then it will have a profound implication for the apparent success of string theory in potentially addressing the black hole information loss problem in classical general relativity, because the problem will have literally evaporated. This will not mean that the string accounting of p-brane states is incorrect, just that string theory would not have been needed to solve this fundamental problem that otherwise appeared to suggest the need to move beyond general relativity. String theorists will have to turn their attention to other problems the theory might more uniquely address.

Which brings us back, finally, to Einstein's revenge: the cosmological constant problem. This, after all, remains the key mystery in theoretical physics, and the clearest place where a theory of quantum gravity should shed some light. And it is the place where, I think it is fair to say, string theory had its biggest unmitigated lack of success. Nothing in all the work following the first string revolution, or even immediately following the discovery of the importance of D-branes and the emergence of M-theory, had shed any light on the question of how the energy of empty space could be precisely zero.

So, when in 1998 cosmological observations led to the discovery that the energy of empty space isn't precisely zero, just almost zero, everyone–including string theorists–stood up and took notice. Maybe, just maybe, this finding might provide a vital clue that could either vindicate the string revolution or help us move beyond it.

The result was a sudden new explosion of interest in–you guessed it–extra dimensions–but not the hypothetical, aetherial, and perhaps illusory extra dimensions that had so fixated the ten- or eleven-dimensional imaginations of string theorists. Rather, they were concrete and even potentially accessible extra dimensions that might literally be hiding behind the looking glass or on the other side of the wardrobe.

CHAPTER 16
D IS FOR *BRANEWORLD*

The small man said to the other:
"Where does a wise man hide a pebble?"
And the tall man answered in a low voice:
"On the beach."

–G. K. Chesterton

It is easy, in the midst of discussing such things as D-branes and supersymmetric state counting, to forget precisely what we are really talking about here. In order to understand what might otherwise be considered a somewhat esoteric corner of physical theory–the intersection of gravity and quantum mechanics–string theory or its successor, M-theory, suggests that we need to believe that the world of our experience is but a minor reflection of a higher-dimensional reality. The tragedies of human existence may be very poignant, and the evolution of our visible universe may be remarkable, but actually, they are all fundamentally a cosmic afterthought.

Somehow the key to our existence lies in the poorly understood, but remarkably rich, possibilities available to a universe with perhaps seven extra dimensions, although one or more of these may not behave like any dimensions we have experienced. Moreover, the conventional wisdom, steeped in a tradition established by Kaluza and Klein almost a century ago, suggests these seven dimensions are "compactified," bundled up for as-of-yet unknown reasons into regions so small that a pebble lying on a beach would be, by comparison, as large as our own galaxy is compared to the pebble.

At the same time deeper questions arise, some of which I have already considered. If the convolutions appropriate to extra-dimensional physics really do ultimately lead to a picture of the four-dimensional universe that accurately resembles the reality we experience, but if at the same time these dimensions remain forever hidden, ephemeral theoretical entities, inaccessible to our experiments, if not our imagination, then in what sense are these extra dimensions more than merely mathematical constructs? What, in this case, does it mean to be *real*?

There are times when I have wondered whether Michael Faraday, as he developed his fantastic mental images of hypothetical electric and magnetic fields permeating space in 1840 felt that they were so simple and beautiful that they had to exist. Or did he consider them to be merely a convenient crutch, so that someone like himself, unschooled in mathematics, could comprehend in an intuitive way some sliver of the physical world?

As I have mentioned, there is, of course, a noble tradition in physics of mathematical crutches turning out to have a physical reality. Faraday's electromagnetic fields are just one example. Quarks, when they were first introduced, were also seen primarily as a mathematical classification scheme, rather than as real entities. So, too, were atoms, for that matter. Indeed, Ludwig Boltzmann committed suicide in part because he felt he could not convince his contemporaries that atoms had to be real.

On the other hand, many mathematical models that have been proposed have thus far borne no relation to the real world–even mathematics that at one time or another seemed to show great promise. So the questions posed earlier remain relevant, and short of a theoretical breakthrough that unambiguously allows a prediction of unique laws of nature that match the ones we observe, the only way we may know if any of these higher-dimensional imaginings are correct is if somehow we can ultimately experimentally probe the extra dimensions, either directly or indirectly.

Traditionally in string theory this has seemed like a colossal long shot. If the string scale is comparable with the Planck scale–about 10^{-33} cm, where quantum mechanical effects in gravity are presumed to become important–it is as far removed from anything remotely accessible in the laboratory.

Imagine you were looking at our galaxy through a distant telescope from another galaxy far, far away. Say your telescope could just barely resolve individual stars in the Milky Way, as the Hubble Space Telescope can in the nearby Andromeda galaxy, two million light-years away. The problem of measuring extra dimensions on the Planck scale is for us, then, similar to the problem of your trying to detect and probe individual atoms in that distant galaxy using your telescope!

The past decade has, however, produced some remarkable transformations in the way we think about fundamental physics, driven largely, I am happy to say, by the surprises nature has wrought.

Nature provided a cosmic wake-up call, in the form of dark energy, that not even those fully immersed in eleven-dimensional mathematics could ignore. In particular, the discovery that dark energy dominates the expansion of the universe is so shocking that it seems very likely that it is related to something fundamentally profound about the structure of space and time. And since string theory has taken as its mantra the revelation of profound new truths in these areas, the unexpected appearance of dark energy cried out for attention. Or, at the very least, it was irritating to the point of distraction.

The distraction was key, however. It stood as a stark reminder that, at the earliest moments of the big bang, what is now the visible universe was of a size comparable to the microscopically small scale of the purported extra dimensions. Thus perhaps the universe itself could provide the experiment that might ultimately reveal these extra dimensions for all to witness.

I remember David Gross's telling me in 2002 why string theorists had suddenly become so interested in cosmology. The big bang, taken back to $t=0$ inevitably leads to a singularity (a point of infinite density) at the beginning of time. There is clearly something physically implausible about such a state of infinitely high density. One of the main virtues of string theory, however, is its apparent ability to dispense with such infinite singularities, at least those that seemed to plague general relativity. Thus, string theory might be able to dispense with the big bang singularity, and perhaps in the process explain the mystery of dark energy. I confess that in a skeptical moment I responded to David by expressing the concern that string theory might instead do for observational cosmology what it has thus far done for experimental elementary particle physics: namely, nothing.

Sarcasm aside, however, in 1998 several theoretical breakthroughs transformed the way much of modern research is being performed, and have made the question of the possible reality of extra dimensions something of immediate and practical interest. They did not arise from cosmology, however, although they opened up, literally, a whole new universe of cosmological possibilities. Rather, they were inspired by a new consideration, reflected in the glow of D-branes, of the very same problem that first motivated many particle physicists to adopt supersymmetry as a useful guiding principle in nature: the hierarchy problem.

Recall that the hierarchy problem in particle physics relates to the question of why the GUT energy scale, where the three nongravitational forces in nature may be unified, could be fifteen orders of magnitude larger than the scale at which the weak and electromagnetic interactions are unified. Worse still, the Planck energy scale, where quantum gravity should become important, is seventeen orders of magnitude larger than this latter scale. Not only are these large discrepancies of scale inexplicable, but it turns out that formally, within the context of the standard model of particle physics without supersymmetry, this hierarchy is unstable. Namely, as I have described, the effects of high-energy virtual particles will tend to lead to intolerably large corrections in the low-energy theory.

In 1998 physicists Nima Arkani-Hamed, Savas Dimopoulos, and Gia Dvali proposed a dramatic new way of avoiding the hierarchy between the Planck scale and the weak scale. They suggested that perhaps the Planck scale is not really where we think it is. The group was motivated by considering the possible existence of extra dimensions, and also indirectly by the development of D-branes in string theory. As I shall describe, their argument relied on the possibility that perhaps the extra dimensions, or at least one of them, might in fact not be microscopically small, but rather could be "almost" visible–perhaps, in fact, the size of a small pebble lying on a gravel road.

An immediate question that comes to mind when this possibility is raised is: If the extra dimensions are that big, why don't we see them? A possible answer lies in the magic of D-branes. Remember that in string theory, open strings can end on D-branes, so that the charges on the ends of these strings, and the Yang-Mills fields and forces associated with these charges, might reside only on the D-branes. Remember, however, that gravitons, the particles associated with gravitational fields, are associated with closed string loops (i.e., objects without ends). These loops can also move about in the space between the branes, and thus gravitational fields are not restricted to exist only on the D-branes, but can also exist in the "bulk," as the space between the branes is called.

Thus, imagine that the three spatial dimensions of our experience lie on a three-brane "surface" in a higher-dimensional space. If gravity is the only force that can exist outside of our three-brane, then only gravity could probe these extra dimensions.

How would gravity do so? Well, Newton's theory of gravity tells us that the gravitational force between two objects falls off inversely as the square of the distance between them. This is, after all, precisely the same behavior that characterizes the electric force between charged objects.

We now return at long last to Michael Faraday, whose brilliant idea of field lines helped to provide an intuitive understanding of why electric forces actually fell off as the inverse square of distance. Remember that if field lines move out in all directions from a charged particle, the number of field lines per unit area crossing any surface will fall off inversely with this area or, equivalently, inversely as the square of the distance from the source.

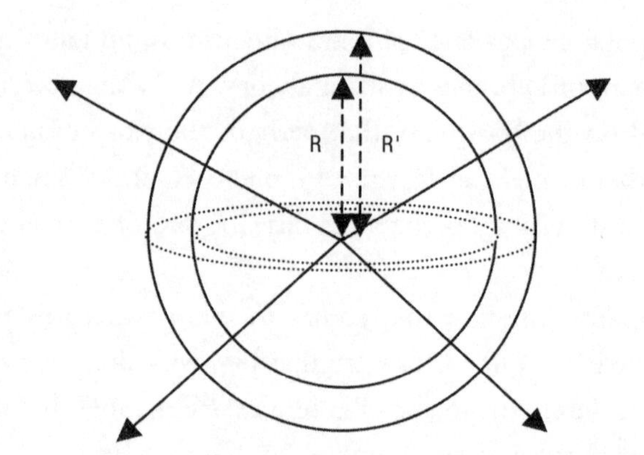

THE ELECTRIC FORCE FALLS AS R^{-2} IN FARADAY'S PICTURE BECAUSE THE FORCE IS PROPORTIONAL TO THE NUMBER OF FIELD LINES PER UNIT AREA CROSSING A SURFACE. IF FOUR FIELD LINES ARE LEAVING THE CHARGE AT THE CENTER OF THE SPHERE ABOVE, THEN THE FORCE AT A DISTANCE R IS PROPORTIONAL TO $4/R^2$ AND AT A LARGER DISTANCE IT IS PROPORTIONAL TO $4/R'^2$.

What Arkani-Hamed and his collaborators proposed was that a similar argument would suggest that if gravitational forces propagated in extra dimensions, as well as in our three-dimensional space, then the strength of the gravitational force measured between massive particles in our space would fall off faster than the inverse of the square of the distance between them. Imagine, for example, a single extra dimension. If field lines could spread out in our three dimensions, but also in this extra dimension, then the number of field lines per unit area would fall off as the area of a three-dimensional spherical surface (bounding a four-dimensional volume), and not as the two-dimensional spherical surface bounding a three-dimensional volume that we normally picture when we draw field lines spreading out into space. Since the area of a three-dimensional spherical surface increases with the cube of its radius, and not the square of its radius, as in a two-dimensional spherical surface, this means that the strength of gravity would fall off inversely with the *cube* of distance, not the *square* of distance.

There is, of course, a slight problem here. Newton achieved fame and fortune by demonstrating that a universal gravitational force that fell off with the square of distance could explain everything from falling apples to the orbits of planets! So, what gives?

Well, it is true that gravity has been measured with great precision to have an inverse square law on scales ranging from human scales to galactic scales. But, as Arkani-Hamed and collaborators pointed out, it hadn't been so measured at scales smaller than about a millimeter.

Imagine, then, that the extra spatial dimension has a size of a millimeter. Then for objects separated in our space by less than about a millimeter, the force of gravity will fall off with the cube of distance. But once objects get separated by a larger amount than this, the gravitational field lines from one particle cannot spread out any more in the extra dimension as they can do in our three dimensions of space. As the field lines can continue to spread out only in the three remaining large dimensions on scales larger than a millimeter, the gravitational field again begins to now fall off inversely with the square of distance. An example starting with three dimensions, one of which is small, is shown below:

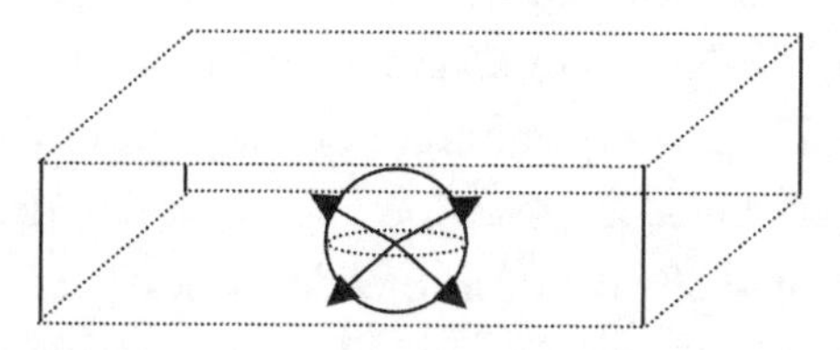

IN A THREE-DIMENSIONAL SPACE OF FINITE EXTENT IN THE VERTICAL DIRECTION, THE FIELD LINES WILL SPREAD OUT AS $1/R^2$ FOR DISTANCES SMALLER THAN THE SIZE OF THE VERTICAL DIMENSIONS.

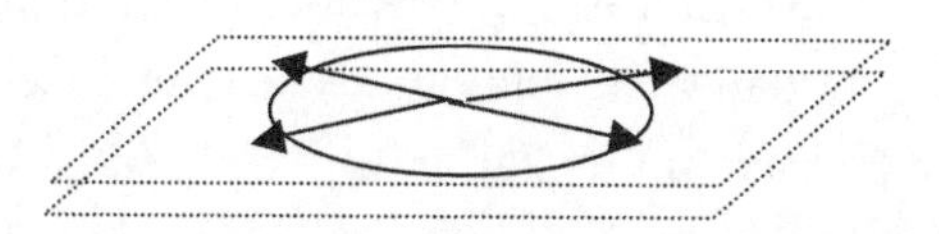

FOR DISTANCES MUCH LARGER THAN THE EXTENT IN THE FINITE VERTICAL DIMENSION, THE FIELD LINES CAN EFFECTIVELY ONLY SPREAD OUT IN TWO DIMENSIONS, SO THE FIELD WILL FALL OFF AS $1/R$.

What is the net effect of all of this? Well, if the size of the extra dimension is *R,* then gravity falls off with one extra power of distance between the Planck scale and *R,* compared to what it would be if there were only three dimensions for gravity to propagate in. Thus, by the time objects are separated by a distance *R,* the gravitational attraction between them would be weaker by a much larger factor than they would otherwise have been. If one is only measuring gravity on scales larger than *R,* gravity would then be measured to fall off with an inverse square law, just as Newton argued, but by this time the apparent strength of gravity would be much smaller than it would have been if the gravitational field had not been able to fall, at least temporarily, faster than inversely with the square of distance.

Without knowledge of these extra dimensions, this extra suppression factor would simply be incorporated into the basic definition of the strength of gravity itself. This strength is given by what we conventionally call Newton's constant, which appears in the formula for the inverse square law gravitational force between two bodies.

By a similar argument, if there are two extra hidden compactified dimensions, then gravity will fall off by an even greater factor between the real Planck scale and the scale, *R,* of the extra dimensions, and so on for yet more compactified dimensions.

Because it tells us how strong the gravitational force is between measured bodies, we use Newton's constant to determine on what distance or energy scale we expect that quantum mechanical gravitational effects should become significant. Thus, it is Newton's constant that determines the value of what we conventionally define as the Planck scale. As a result, because of the hidden effects of the extra dimensions on scales between the Planck scale *R,* we would "inaccurately" deduce the Planck scale to be much higher than it actually is. This is because the strength of the gravitational force would grow much faster as one decreases the distance between massive objects, on scales smaller than *R,* than we would otherwise have expected.

What Arkani-Hamed and collaborators realized is that this would allow the possibility that the real Planck scale might actually be equal to the distance scale at when the weak and electromagnetic interactions are unified, instead of seventeen orders of magnitude smaller, as we would otherwise estimate based on our incorrect extrapolation of the behavior of gravity on scales smaller than *R.* This would therefore naturally explain the apparent large hierarchy between the electroweak scale and the Planck scale. The hierarchy problem would therefore be a problem of our own making; no such actual hierarchy would exist in nature.

There is an immediate concern, however. With just one extra dimension, the extra falloff in the strength of gravity from the real Planck scale to the size of the extra dimension, *R,* is sufficiently slow so that to yield the strength of gravity that we actually do measure on large scales forces this latter size *R* to be roughly equal to size of our solar system! This is clearly impossible, since this would mean that gravity would be measured to fall off inversely with the cube of distance throughout our solar system, when it was precisely the measurements of gravity over our solar system that led Newton to propose the inverse square law in the first place.

Clearly then, one such extra large dimension is ruled out by observation. However, if there were two extra new dimensions, then because gravity would fall off even faster with distance for distances smaller than *R,* this would allow the size of the extra dimension, *R,* to be much smaller than it would be in the case of one extra dimension. If one works out the numbers, the scale *R* would only have to be about one millimeter. In this case, the extra dimensions would indeed be as large as small pebbles.

Not only is such a possibility not ruled out, but this is precisely the scale at which new experiments had been designed to explore the inverse square law behavior of gravity. If Arkani-Hamed and collaborators were correct, and if there are only two extra large dimensions, these experiments could measure something quite different, revealing for the first time the hidden dimensions that have otherwise remained within the realm of theorists' imaginations for all these years.

To recap: It is perfectly possible for extra dimensions to exist and be relatively large, provided that the only force that can propagate in these extra dimensions is gravity. Moreover, one can resolve the hierarchy problem for gravity, making the Planck scale, and with it presumably the string scale, essentially equal to the electroweak scale if there are two extra large dimensions into which gravity can propagate, both of which are about a millimeter in size.

As I have indicated, the possibility that extra dimensions, such as those that might be associated with string theory, might be large enough to actually be measured sent a jolt of excitement through much of the particle physics community that was perhaps stronger than any that had been experienced since the first string revolution of 1984. Suddenly a host of potential new experimental probes–not only of quantum gravity, but also of string theory and even extra dimensions–would become feasible.

One of the most exciting such exotic probes involves exploring strings and extra dimensions at current or planned particle physics accelerators.

For if the Planck scale and with it the string scale coincide with the electroweak scale, then machines designed to explore weak interaction physics could uncover exotic new phenomena. Higher-energy string excitations in extra dimensions might be excited in high-energy particle collisions, which would be manifested in precisely the same tower of new particle states (albeit at now much higher energies) that had first been predicted when strings had been proposed as a theory of the strong interactions. Equally interesting would be the possibility that some of the energy in these high-energy collisions might literally disappear in gravitational waves that could move off into the extra dimensions.

Finally, perhaps the most exciting prediction of all would be that gravity itself would become strong enough at the electroweak scale so that new quantum gravitational phenomena might be directly observable there. For example, high-energy collisions in new accelerators might produce primordial, elementary, particlelike "black holes," which might then spontaneously decay in a burst of radiation, as predicted by Hawking. Not only would such new signatures be striking, they would allow us to confirm one of the key phenomena predicted to occur when quantum mechanical effects are incorporated into gravity and ultimately directly explore one of our most puzzling paradoxes, the information loss paradox.

Any of these experiments might be exciting enough to get one's juices flowing, even if they are long shots, but for those who truly crave dimensions large enough to hide aliens in, millimeter sizes, even if huge by comparison to what had previously been assumed in string theory, just don't cut it.

Happily an even more exotic possibility was independently proposed within a year of Arkani-Hamed and coworkers' theory, by Lisa Randall, now at Harvard University, and a past student of mine, Raman Sundrum, currently at Johns Hopkins.

Randall and Sundrum argued that there is another way to resolve the hierarchy problem using extra dimensions that is quite distinct, and certainly more subtle, than the mechanism proposed by Arkani-Hamed and colleagues. They proposed starting with a single compact extra dimension, but not one completely independent of our own. In the true spirit of *Star*

Trek they introduced what they called a "warp factor," though theirs has nothing to do with faster-than-light travel. Rather, it arises from the suggestion that an extra dimension exists that is strongly curved (or "warped") as one moves away from the three-brane that makes up the three-dimensional world we experience.

What Randall and Sundrum realized is that, in this case, even if the size of the extra compactified dimension is perhaps only of order of ten to fifty times larger than the Planck scale, it is still possible to produce a natural large hierarchy, of perhaps fifteen orders of magnitude, between this scale and the scale of the elementary particle masses and interactions we observe.

The effect is subtle and somewhat difficult to directly picture physically without recourse to mathematics, as it due to effects of curvature in the extra dimension. Remember that general relativity tells us that the curvature of space is related to the overall magnitude of the mass and energy of objects within the space. Now the curvature associated with the warping of the extra dimension near our three-brane, in the Randall-Sundrum picture, could be rather large, characteristic of energies near the Planck energy scale. But, if the extra dimension is perhaps fifty times larger than the characteristic scale over which it curves, then when one solves the full five-dimensional equations associated with general relativity, a hierarchy appears. It turns out that even if, in the five-dimensional theory, the fundamental mass and energy parameters are all of the order of the Planck scale, in our four-dimensional world all fundamental particle masses will instead appear to be suppressed compared to the Planck scale by a factor of 10^{15}.

Randall and Sundrum also showed that there was another slightly more intuitive way of thinking of this problem, in terms of the relative strength of the forces in nature. The hierarchy problem can be recast as follows: Gravity is measured to be more than a billion billion billion billion times weaker than electromagnetism, and even weaker still when compared to the strong force. It may not seem so weak, especially in the morning when you try and pull yourself out of bed, but remember that you are feeling the gravitational force of the entire earth acting on you. By contrast, even a small excess of electric charge on an object such as a balloon produces a large enough electric field to hold it up on a wall against the gravitational pull of the entire earth. The hierarchy problem involves the question of why there is this huge discrepancy.

In the Randall-Sundrum scheme, the warping of space near our three-brane implies that gravity near our brane acts effectively much more weakly than it does outside our brane. The exponential warping, in fact, makes gravity appear exponentially weaker on our brane than it is at the other side of five-dimensional space. If we happened to live on a three-brane located there, which I remind you is located merely a microscopic distance "away" from our world in the extra spatial dimension, gravity would appear to have the same strength as the other forces in nature. The observed hierarchy in our world then becomes merely an environmental accident. Gravity "leaks" into our dimension as surely as Buckaroo Banzai's extra-dimensional nemesis does.

Like the Arkani and coworkers scenario, Randall-Sundrum's extra-dimensional solution of the hierarchy problem would bring extra dimensions into the realm of the testable. In this case, only the massless particle (the graviton) that conveys the gravitational force would be weakly coupled on our brane. As in all compactified theories, there would also be a tower of higher-mass particles that could be produced if one had sufficient energy. In the Randall-Sundrum model, however, these higher-dimensional gravitational modes would have masses characteristic of the electroweak scale and coupling strengths not characteristic of gravity, but rather of electroweak physics. The new extra-dimensional states would thus be produced in great abundance, with observable decay modes, just like ordinary particles, if one had an accelerator that could achieve the necessary energies. And, interestingly, just such an accelerator is being built at CERN (the European Center for Nuclear Research, in Geneva, Switzerland) and is due to come online in 2007 or 2008.

In fact, not only would new higher-dimensional gravitational excitations be produced at such an accelerator if this idea is correct, but at slightly higher scales fundamental strings could also be generated and explored. All the mysteries of string theory or M-theory would be laid bare for experimentalists to probe, even if theorists have remained, by that time, unable to untangle their complexities.

Now, before you go out and buy CERN futures, you might want to step back and note a few of the hidden, but profound, problems with this model as it stands. First and foremost is an issue that has plagued all Kaluza-Klein theories since their origin: Why are the extra dimensions small, and our three-brane possibly infinitely large? There has simply been no good answer to this question in the past ninety years. While a great deal of work has been devoted to trying to find physical mechanisms that would allow such a possibility, no real progress has been made. It is simply assumed that something happens so that the extra dimensions remain hidden, whether they curl up on the size of the Planck scale or are as large as a pebble on the road.

Actually, the situation is often even worse than I have thus far described. In general, it turns out that the dynamic equations of the theory tend to drive the size of the extra dimension to be infinitely large, as presumably are the three spatial dimensions in which we live, even if one initially starts the extra dimensions off to be the Planck scale, say. This embarrassment is solved in the way other similar confusing aspects of string theory and M-theory are sometimes dealt with: Namely, it is assumed that when we fully understand the ultimate theory, everything will become clear.

Nevertheless, aside from this "minor" inadequacy, you may recall that I promised you a theory with a really large extra dimension, not the puny compactified extra dimension that Randall and Sundrum proposed. Happily, for those who find the unbridled optimism of the last paragraph less than convincing, these researchers discovered, within a month of their original suggestion, that a compactified extra dimension was, in fact, completely unnecessary in their warped five-dimensional space-time model. If the space outside our local three-brane was warped, they discovered that the size of the extra dimension(s) could in fact be infinite . . . namely, just as big as the three dimensions of our experience!

To reach this conclusion, they considered–instead of a five-dimensional space with a compactified extra dimension–an infinitely large five-dimensional space with two three-branes located a very small distance apart. The finite volume between the branes mimicked the compactified space of their original model. They then assumed that we live on one of the two three-branes, and considered what happens as one slowly increases the separation of the two branes.

You will recall that in their original scenario, the warping of space-time near our three-brane caused the strength of gravity to fall off exponentially as one approached our brane from anywhere else in the space. In this case, they switched things around, having the strength of gravity fall off exponentially *away* from our brane. One finds, accordingly, that the force of gravity is effectively tied to our brane and, since gravity is our only probe of the extra dimension, even in the limit that the other brane is removed to infinite distances in the fifth dimension the effects of this large extra dimension are completely hidden. Also, it turns out that the masses of the tower of states that occur in higher dimensional theories all tend toward zero, just like the (zero-mass) graviton that is responsible for conveying the gravitational force in our three-dimensional space. But, fortunately all these extra states essentially decouple from matter in our space, and therefore cannot be produced in any present or planned accelerators, and remain completely unobservable. Moreover, any corrections they might provide to the nature of the gravitational force between test particles is suppressed by the ratio of the distance between the test particles divided by the curvature scale in the extra dimension. If this latter quantity is on the order of the Planck scale, then the effect will be unobservably small in any conceivable experiment that might be performed in the future.

What this second Randall-Sundrum model demonstrates is that the conventional wisdom about extra dimensions, stretching back all the way to Kaluza-Klein, was wrong. It is completely possible to hide behind the mirror not only a microscopic extra dimension, as originally envisaged, or even merely a tiny extra dimension, as Arkani-Hamed and colleagues envisaged, but also an *infinite* extra dimension, which would exist in concert with our own three-dimensional space.

Beyond this, what the models of Arkani-Hamed and coworkers and Randall-Sundrum have shown is that if our three-dimensional universe comprises a three-brane within a higher-dimensional space, then it might be possible to resolve at least one fundamental mystery in particle physics while providing a possible new set of signatures that could open up both the extra dimensions, and the complexities of string theory or M-theory to the bright light of experiment.

That is the good news. Once again, however, just below the surface in these models lies a host of problems that suggest, that as far as the possible existence of extra dimensions are concerned, it is very difficult to have one's cake and eat it, too.

First, note that in the second Randall-Sundrum model, which switches branes around from their earlier compactified model, the whole extra-dimensional solution of the hierarchy problem disappears. In the first model the exponential falloff of gravity near our brane is sufficient to make gravity anomalously weak compared to the other forces in nature, while in their second model, unless one fine-tunes things, the four-dimensional Planck scale is identical to the fundamental curvature scale in the higher dimension, which is presumably related to the string scale in this higher-dimensional space. All of the scales are vastly different than the electroweak scale. One must then find another mechanism to enforce the wide disparity between the strength of gravity and the other observed forces in our world. This is inconvenient and, frankly, reduces the motivation for introducing an extra dimension in the first place. Were it not for the fact that three-branes and extra dimensions arise within the context of string/M-theory, one might wonder what one would gain from this albeit fascinating mathematical construction.

But there are other, more fundamental concerns. The key to all of these interesting recent results is the newly recognized possible existence within string theory of three-branes, onto which all nongravitational charges and fields could be constrained, and a higher-dimensional bulk space into which gravity can propagate. However, string theory appears to naturally incorporate branes of all dimensions up to perhaps ten dimensions itself, with all of these comprising all possible orientations within the context of the complicated and as-of-yet not understood ground state of string/M-theory. The notion that our world should lie completely within an isolated three-brane is quite frankly not suggested by anything that is known about string theory at the present time.

From my point of view there is another more immediate issue that strongly diminishes the beauty of the proposed extra-dimensional solutions of the hierarchy problem. The one profoundly important experimental fact we know about the fundamental forces of nature on the scales that we can probe them is that as these scales become smaller and smaller, the strengths of the forces appears to approach a common value. It was this fact that provided one of the most direct pieces of evidence suggesting the existence of a possible grand unified theory in the first place, and that still provides one of the best reasons to believe in supersymmetry as a symmetry of nature. Recall that this symmetry is really the underpinning of all of modern string theory.

But remember, too, that the scale at which this evidence suggests the forces may unify is unambiguously fourteen to fifteen orders of magnitude smaller than the scale at which the weak and electromagnetic interactions themselves unify, and within a few orders of magnitude of the Planck scale itself. Moreover, the other new great discovery of the past twenty-five years in particle physics is the remarkable fact that neutrinos, the ghostly particles that experience only the weak force, are not absolutely massless, but rather have a very small mass, more than hundred thousands times smaller than that of the next lightest particle, the electron. Such masses are not explicable within the context of the standard model, which incorporates all known physics up to the electroweak scale. However, if one adds new physics at the grand unified scale, one can naturally arrive at neutrino masses in this range.

Thus, if one makes the electroweak scale the fundamental scale in nature on which extra dimensions, gravity, and string phenomena arise, one might remove the hierarchy between our four-dimensional Planck scale and the electroweak scale, but in doing so one swims strongly against the tide of experimental evidence. This is not a good precedent for what is supposed to be an empirical science.

It could be that the apparent unification of the strengths of the known nongravitational forces, and the existence of neutrino masses, are just coincidences with no fundamental explanation in terms of grand unification near the Planck scale. But here I paraphrase Einstein: Nature may be subtle, but she is not malicious. If the only evidence that nature seems to be providing us about fundamental scales turns out to be a red herring, this would break a tradition that has stood us in good stead for over four hundred years.

Finally, we once again return to the Achilles heel of all theories of quantum gravity: Einstein's cosmological constant, the energy of empty space. It turns out that in order for our three-dimensional space to exist as a flat three-brane within a warped higher-dimensional space, the vacuum energy associated with the higher-dimensional space would have to be very large, and negative. It would then have to be precisely cancelled on our brane by a contribution that is large and positive in order for the observed energy of three-dimensional space to be both very small and nonzero. In short, the biggest fine-tuning problem known in nature becomes even more significant in these models, which, after all, were motivated by a desire to solve a much less severe numerical issue.

Still, I come here not to bury these new ideas about extra dimensions but to praise them. For all their potential weaknesses, they have revealed as at least experimentally allowable a whole host of possible extra dimensions that had hitherto been considered ruled out. And whenever new theoretical possibilities exist, there is always the chance that nature will actually take advantage of them.

All the problems and challenges aside, the realization that the world of our experience could, in principle, be embedded in a larger space that could become directly experimentally accessible in the near future has caused a tremendous explosion of energy devoted to exploring all the potential consequences of (possibly large) extra dimensions and the branes that may exist within them.

The sociology of physics is a strange and wonderful thing. The reaction to the Arkani-Hamed and coworkers and Randall-Sundrum papers was nothing short of phenomenal. Within six years, no fewer than 2,500 separate scientific papers appeared exploring their ramifications. Like a well-timed drama that somehow captures the public's imagination, the notions of large extra dimensions and/or a low-energy string scale seemed to have everything going for them in the theoretical physics community. They were novel, sexy, and potentially testable.

New phenomena associated with strings and extra dimensions that had previously been assumed to be forever inaccessible are, if these ideas are correct (and I remain dubious), possibly on the verge of being measured in the laboratory. Direct probes of gravity on scales smaller than one mm are being developed that might probe for a change from the inverse square law. Alternatively, if the Planck scale coincides with the electroweak scale, then because we can probe the latter scale with modern particle accelerators, perhaps we could directly use these devices to probe extra-dimensional quantum gravitational phenomena.

But as interesting to physicists as these direct tests might be, the poetic and philosophical implications of potentially large extra dimensions lie elsewhere. Other branes could represent possibly infinitely large alternate universes that could exist, literally, less than a fingernail's width away from our own. Each of these universes could have laws of physics that might be dramatically different from our own as well as a dramatically different life history. And so, even if possible extra dimensions continue to elude the able probes of direct laboratory experiments, it could be that observations associated with the origin and evolution of our entire universe may unlock the door to their discovery.

Evidence for the existence or absence of extra dimensions is likely to come ultimately not from an attempt to understand the dynamics of objects within our universe, but rather from an attempt to understand the dynamics of our universe itself and to address the ultimate questions that have beset science since it first emerged from the fog of history: How did the universe begin? How will it end?

And it is here, as I earlier suggested, that string theory, too, must ultimately face the music. If it is really ever to provide an explanation of anything we see, much less everything we see, it must address the fundamental nature of that which we cannot see but which we know is there. It must explain the energy of nothing.

CHAPTER 17

A THEORY OF NOTHING?

Wherever you go, there you are.
—*The Adventures of Buckaroo Banzai across the Eighth Dimension*

What could be more romantic than the notion that extra dimensions might not be truly hidden, but that objects from our universe might cross over into this new realm? And since physics is a two-way street, with that possibility comes a more exciting or perhaps terrifying one: What if material or information from these extra dimensions can "leak" into our own world? What if, ultimately, the source of our own existence lies across that invisible boundary?

As we have seen, these questions have been the fodder for speculation and belief for almost four centuries, since sixteenth-century theologians first speculated that spirits and angels emerge from the extra-dimensional universe. But now they have reemerged in a new scientific context that might actually be testable.

For a literary mind, the science fiction possibilities of these concepts are endless, and Buckaroo Banzai's adventures are merely one particularly wacky manifestation. So, too, for physicists and their graduate students, long starved of new calculations that might be performed and even tested, the possibility of large extra dimensions and the existence of other branes than our own have provided countless new opportunities for exploration and creative expression. These have become popularly known as "Braneworld" scenarios, which sounds like a science fiction movie title as much as anything ever did. Even Stephen Hawking has gotten into the act with a recent popular lecture entitled "Brane New World."

In some sense it is appropriate that this research area does sound like science fiction, because most of it probably is. What is too often underappreciated about science is that almost all of the ideas it proposes turn out to be wrong. If they weren't, the line between science and science fiction would be much less firm. But the "present" can perhaps be defined as that time when we teeter on the edge of understanding, and where the line between speculative science and science fiction is most easily blurred. And that is precisely where we now are in this narrative of our ongoing love affair with extra dimensions.

This is not to suggest, however, that all ideas are equally attractive. Over the past five years, hundreds, if not thousands, of scientific papers have been written considering cosmological possibilities that might be associated with Braneworld scenarios. One cannot do justice to all of them, but the greatest justice I could probably do to many of them is to not mention them here.

Nevertheless, it is undeniable that the mere fact that we might live on a three-brane in a possibly infinite or large but compactified extra-dimensional space dramatically has broadened the scope of cosmological investigation. For example: What may have caused our three spatial dimensions to have become potentially so much larger than the other extra dimensions, and could the latter's dynamic evolution have an impact upon the cosmological evolution of our visible universe? What about the possible existence of other nearby branes? In the earliest moments of our big bang expansion, when the scale of our presently observable universe was as small or smaller than the present size of any compactified extra dimension, how could the presence of significant other dimensions have affected both the origin and evolution of our universe? And, how might our brane evolve dynamically within the bulk space today, or, equivalently, how might the changing nature of gravity on small or large scales in extra-dimensional Braneworld scenarios have an impact upon current measurements in observational cosmology?

The first question has been around in one form or another since Kaluza and Klein first wrote down their ideas involving compact extra dimensions, and, as I have argued earlier, it is fair to say that no very good answer has yet been provided. If one compactifies extra dimensions into some small radii, *r*, then the size of these radii leaves an imprint on the remaining large dimensions via the existence of new fields in nature, called moduli fields. String theory is replete with such moduli fields. One can explore the dynamics of these fields and it turns out that they tend to want to relax to a zero value, which, in the higher dimensional picture, corresponds to the radii of the extra dimensions going to infinity. To stop this runaway expansion of dimensions, one generally has to introduce ad hoc mechanisms, which is one of the reasons that the Randall-Sundrum warped-extra-dimensional scenario, with its infinitely large extra dimensions, was proposed. Nevertheless, there have been suggestions that somehow the expansion of our three dimensions might arise at a cost to the extra dimensions, with our dimensions expanding, while the others perhaps contract. While this notion has some aesthetic appeal, no otherwise attractive mechanism has been proposed to generate a workable model.

The next question, regarding the possible existence of other branes, and their potential effects on our own, is more intriguing. One particularly inventive proposal in this regard actually explored the possibility that these "extra" branes might actually *be* our own.

Shortly after the first Arkani-Hamed and colleagues (ADD) proposal for large extra dimensions, these authors, along with several others, proposed that our brane might actually be folded over on itself many times, with different sheets located less than a millimeter away in the extra compact dimension. Since electromagnetic radiation and all nongravitational fields propagate only along our brane and not out into this extra dimension, these other regions would be invisible to us as long as the "folds" in our brane occurred at distances along our brane so far away that light has not yet had sufficient time since the big bang to travel across such distances. Thus, the only effect of these extra sheets would be their gravitational effects on us, since gravity can cross into the bulk between them.

But since these extra sheets are really part of our brane, the laws of physics on them are identical to our own. Thus, otherwise invisible gas, stars, and galaxies could exist superimposed "on top" of our space. The authors of papers on this topic have suggested that these invisible objects might somehow comprise the dark matter that we infer to dominate the mass of our galaxies, for example.

While this might be plausible in a science fiction universe, it will not pass muster in the real universe. The question of why these invisible stars and galaxies should tend to cluster along with our own but why the material in them should nevertheless spread out in halos around visible galaxies, was unanswered. Indeed, there are a host of other issues that must be addressed, including what mechanism might fold our brane and keep it folded.

The problems that beset this idea are typical of many Braneworld scenarios for cosmology. The freedom allowed by extra dimensions introduces lots of exotic possibilities, but almost every one of them involves a set of new cosmological problems that must be dealt with in order to agree with observations. Most important of all, though, is the fact that there are often very plausible non-Braneworld approaches that address many of the same cosmological issues these new scenarios propose to deal with. For example, elementary particle physics now offers many realistic candidates for dark matter along with natural mechanisms to explain how it might have survived the earliest moments of the hot big bang so that it might come to dominate the mass of the universe today. Morever, particle physics provides very elegant mechanisms for generating the density perturbations in the very early universe that might ultimately collapse to form galaxies of visible and dark matter. It is not clear that the additional intellectual overhead associated with branes and extra dimensions is needed to explain anything that we might otherwise explain without it.

Another example involves an idea that has become central to modern cosmology, inflation. Recall that in 1980 the physicist Alan Guth proposed that phase transitions in the early universe could lead to periods of rapid early expansion. What he also showed is that such periods would resolve two otherwise completely inexplicable but central features of our universe, including its remarkable isotropy (i.e., uniformity) on large scales and the fact that the universe does not appear to be curved on large scales. Moreover, it was subsequently demonstrated that quantum mechanical processes during inflation could generate density fluctuations that could in principle later gravitationally collapse to produce the observed distribution of galaxies in the universe. Recently the observation of small temperature fluctuations in the cosmic microwave background radiation appears to be completely consistent with this scenario. While such consistency cannot prove inflation actually happened in the early universe, it is strongly suggestive.

Nevertheless, in spite of the beauty of the idea of inflation, no particle physics models have been developed that provide compelling or even particularly attractive mechanisms that might underlie it. One might wonder therefore whether, in this case, Braneworlds might come to the rescue.

Within a year after Arkani-Hamed and colleagues' article, Gia Dvali and his collaborator Henry Tye recognized that as two branes approach each other, the residual moduli field in our dimension that is results from their separation in the extra dimension strongly resembles the kind of field that previously had been proposed, ad hoc, to result in an inflationary phase in the early universe. Furthermore, depending upon the net energy associated with empty space on each of the branes, there would be forces of either attraction or of repulsion between the branes that might produce a period of inflation that could in principle gently end as the two branes approached or diverged from each other.

While this picture has the advantage of allowing an inflationary phase without the need to introduce additional elementary particles and fields in the early universe, it is not without its own weaknesses. The brane energies have to be carefully adjusted for the scenario to work. More than this, it is very difficult in these scenarios, once brane interaction energies are converted into the matter and radiation necessary to produce the early hot universe that was the precursor of the universe we now observe, to stop most of the produced energy from instead being transferred to invisible gravitational modes that would be radiated off into the bulk and not on our brane.

Another imaginative tack has been to use Braneworlds to almost completely avoid the outstanding issues associated with inflationary universe models. Paul Steinhardt at Princeton and Neil Turok at Cambridge have recently proposed using a Braneworld scenario to allow a return to the "cyclic universe" models that were in vogue before the success of the current big bang picture.

They have proposed a model called the ekpyrotic universe. In the ekpyrotic universe the current period of accelerated expansion observed in our space is related to the separation of our brane and another one embedded in some higher dimensions. Ultimately, however, these two branes will stop moving apart and will begin to draw closer together. When this happens, our universe will undergo a collapse and reheat again in a reverse of the current big bang expansion. These two branes will eventually cross through each other, producing a burst of energy that will generate another big bang expansion that might proceed again for billions of years as the two branes once again separate. Ultimately, as the interaction energy between the branes begins to dominate, our brane will once again experience an exponential expansion just before the attraction between the branes once again causes them to stop separating and repeat the whole process.

The interesting aspect of this model, and the part that has a certain science fiction charm, is the fact that the period of inflationary expansion that ultimately causes the universe to look flatter and smoother happens near the end of each big bang expansion phase instead of at the beginning. Thus, the reason our universe looks isotropic is that in the cycle that preceded ours, before all stars, galaxies, and civilizations in that expanding universe were subsequently destroyed in a big crunch preceding our own big bang, astronomers in that doomed universe would have measured their universe to be undergoing an accelerated expansion, just as we are measuring today.

As aesthetically pleasing as such an oscillating universe with no beginning and no end might be, however, in order for it to be viable one must ensure that the isotropic, relatively uniform universe that supposedly results during the final accelerating expansion in one phase can survive the subsequent collapse and collision of the two branes to produce isotropic conditions for the next big bang. It is not at all clear that this is possible. In particular, one must make certain that the two colliding branes are precisely aligned as they collide in order for this picture to be viable. Of greater concern, perhaps, is that, if one asks what the natural period is for this oscillating universe to go through each cycle, the timeframe is of the order of the Planck time, about 10^{-43} seconds! In order to produce universes that expand for at least ten billion years, the parameters of these models must therefore be very carefully fine-tuned.

By now I hope you get the general flavor of the dilemma. Braneworlds provide lots of new possibilities for cosmology and the early universe, but nothing yet to write home about, or at least, it seems to me, nothing that yet seems much more attractive than the theories we already have.

But there remains hope, in the form of the one inexplicable, crazy facet of modern cosmology that so far has resisted all efforts to even begin to understand it: dark energy. The fact that empty space appears to carry an energy that is large enough to dominate the expansion of the universe today, yet is 120 orders of magnitude smaller than what one would expect on the basis of conventional ideas associated with the quantum physics of four dimensions, literally begs for some out-of-this world ideas to explain its existence.

The key problem associated with trying to describe dark energy in terms of fundamental particle physics is that the effect of dark energy is primarily felt at large scales. On the scale of the solar system and smaller, for example, the gravitational forces associated with matter (i.e., the sun and planets) overwhelm the minuscule repulsive effect induced by a small cosmological constant. But on the scale of clusters of galaxies and larger, the repulsive force due to this energy of empty space dominates.

The problem is actually even more serious than this. The focus of our efforts to understand the fundamental laws of nature has involved examining phenomena at ever-smaller scales. When we first began to explore the nature of atoms we discovered the laws of quantum mechanics. Similarly, as we explored the nature of the nucleus, we discovered the weak and strong forces. If all of these get unified in some grand unified model, we expect the new physics might appear on scales much smaller than this. Even those models that place a string theory scale near the electroweak scale predict that if new physics appears, it will be on scales smaller than those that we have currently been able to measure.

Indeed, we now realize precisely how it is possible that new physics can emerge on ever-smaller scales without impacting upon the well-understood explanations of how the universe operates on larger scales. Quantum mechanics, for example, is largely irrelevant when considering the motion of baseballs or cannonballs, which is why Newton didn't have to know about it when he developed his classical laws of motion.

But the problem with trying to understand dark energy from a fundamental physics perspective is that it appears to be a large-scale phenomenon, relevant to the expansion of the entire universe. The actual amount of energy associated with empty space in the room in which you are reading this is incredibly small–so small, in fact, that it is hard to imagine how any revision of the laws of physics that might accommodate it would not also dramatically affect physics on all higher-energy scales. This, in a nutshell, is the fundamental problem that has bedeviled all attempts, including string theory attempts, to resolve the cosmological constant problem on fundamental grounds.

In this regard, a particularly creative and novel use of the Braneworld idea was proposed by Gia Dvali–whose name has already come up several times as one of the most active and inventive young theorists in this area today. Dvali, with his NYU colleagues Gabadadze and Porrati, examined in a series of papers a possibility that was in some sense diametrically opposed to the extra-dimensional scenarios that had been considered previously.

They imagined an infinite-volume extra-dimensional space in which gravity could propagate. They then argued that if one were confined–as we presumably are–to a four-dimensional brane, then under certain circumstances, for relatively short (on a cosmic scale) distances and times, one might actually measure the gravitational interaction between objects on our brane to be that calculated by Newton and Einstein. However, over long times and distances the gravitational fields could "leak" into the extra dimension. The net effect would be to change the nature of gravity at large distances and times, not small ones.

Not being ones to hedge their bets, Dvali and colleagues pointed out that there were two different ways that this kind of mechanism might address both the nature of the dark energy that is apparently driving the observed accelerated expansion of the universe, and the broader and more fundamental cosmological constant problem. It would do this by getting rid of both of them.

As far as the nature of dark energy is concerned, one of the interesting implications of modifying gravity at large scales is that one might modify Einstein's equations in a way that would produce accelerated expansion on sufficiently large scales, even without any dark energy as a source. This is, of course, very attractive, because dark energy isn't.

Nevertheless, even if one were to get rid of the need for dark energy, one still has to explain why its value isn't gigantic. Specifically, we would need to solve the cosmological constant problem by explaining why quantum mechanics doesn't produce a vacuum energy that results in even greater acceleration than we would observe today from these additional new gravitational shenanigans at large distance. Here again, Dvali and colleagues provided at least the germ of an interesting idea.

If, on the largest scales, gravity is really five-dimensional, and not four-dimensional, then the relevant vacuum energy to which gravity would be sensitive is the full vacuum energy in five dimensions. It just might be possible to imagine symmetries, like supersymmetry, that could be exact in five or higher dimensions, while broken in our four-dimensional world. Such symmetries might imply that the higher-dimensional vacuum energy was zero. Thus, even if there existed a nonzero cosmological constant on our brane, it could be that gravity on large scales would not "feel" this cosmological constant.

These ideas are fascinating, in part because they are so heretical and counterintuitive. Unfortunately, however, they are also quite provisional. There are a lot of "mights" in the preceding paragraph, and no real model including all of these features has been developed and explored. What is worse is that this possibility may in fact have already been ruled out by observations.

As Dvali and his colleagues have shown, the presence of such infinite-volume extra dimensions is not completely hidden. Because of the nature of general relativity, it turns out that, in their models, the effect of five dimensions changes gravity slightly on all scales, so there must be small corrections to the Newtonian gravitational attraction between all objects, no matter how small or close. However, very high-precision experiments that have been conducted on our solar system would strongly constrain the magnitude of any such possible corrections to the force between the sun and the inner planets, for example. If one is an optimist, one might hope that as these measurements improve, deviations will be seen that imply that perhaps gravity on large scales really *is* leaking away. In any case, for the moment the upper bounds on what is allowed come very close to the level one might expect from such extra-dimensional effects, but work remains to be done to verify this in detail.

We are thus left at present with the somewhat uncomfortable situation that Braneworlds, for all of their hope and hype, haven't yet demonstrated what it takes to be compelling. Their major virtue at this point in time is that some of their consequences are at least in principle testable, via either cosmological observations or the next generation of particle accelerators.

Which brings us back at long last to string theory, M-theory, and the Theory of Everything. Ultimately we should recall that Braneworld ideas seem at best poor approximations to reality, if string theory is correct. What the notion of large or possibly infinite extra dimensions has done is borrow some of the facets of string theory while ignoring the bulk of the theory (forgive the pun), about which, as I have explained, we currently only have the vaguest notions. It seems to me to be a very big long shot that an apparently ad hoc choice of what to keep and what to ignore will capture the essential physics of our universe. To truly understand the origin and evolution of our universe from its earliest moments, if M-theory really corresponds to reality, we will almost certainly be required to understand that theory better than we currently do.

And as I have now stressed several times, one of the most significant areas where string theory has had no success thus far (amidst a long list), and where it may ultimately rise or fall, is the attempt to understand the energy of empty space. String theory never explained why the vacuum energy should be precisely zero when we thought that was the case in the 1980s and 1990s, nor did it predict that it might be nonzero but unbelievably tiny, as it would seem to be in order to explain current cosmological observations. Braneworld proposals notwithstanding, it is most likely that to understand why empty space appears to gravitate the way it does will require a complete theory that merges quantum mechanics and gravity. At present M-theory/string theory is the only game in town, even if no one yet knows what the rules are.

So, even as the nature of M-theory seems to be increasingly elusive and the likelihood that a higher-dimensional theory will clearly resolve other fundamental questions in particle physics is becoming murky, some string theorists have now turned their attention to this fundamental puzzle in the hopes that cosmology might provide a beacon that has otherwise been lacking that can illuminate these dark and hidden worlds.

This has resulted in yet another fascinating sociological metamorphosis of the theory, with warts becoming beauty marks. The presence of dark energy may have completely changed the landscape of modern cosmology, but string theory was not to be outdone: It has produced its own landscape.

Recall that one of the apparent vagaries of string/M-theory is the fact that even if the underlying symmetry structure and number of dimensions associated with the theory were to become explicitly known, the fundamental nature of physics in our three-dimensional space might nevertheless remain undetermined. This is because in order to reduce the theory from ten or eleven dimensions to four, one generally must compactify the extra dimensions, or at least explain how they might be otherwise unobservable in our space at the present time.

For now, there is no guiding mathematical principle that tells us which compactifications are reasonable. The number of different corresponding possible ground states of our universe corresponds roughly to the number of inequivalent possible compactifications. With ten dimensions to start with, and a host of Calabi-Yau possible compactification manifolds, for example, it has been estimated that there may be more than 10^{100} different possible inequivalent ground state configurations that might describe viable four-dimensional universes, and that might result from a single underlying M-theory.

When this was first realized, it looked like bad news for string theory, because it meant that any hope of predictability might go down the drain. Without any way to choose between different ground states, each of which would represent a four-dimensional universe that might have a different configuration of forces and underlying symmetries and a completely different spectrum of elementary particles, the long-sought uniqueness of string theory seemed ephemeral at best. For some time this final step, compactification, was frankly not emphasized in discussions heralding the beauty of string theory.

But with an exceedingly small vacuum energy apparently present in our universe, suddenly the terms have changed. The plethora of possible ground states of the theory, and the nonuniqueness of string-theoretic predictions, have become a virtue, offering hope where none had appeared before.

The source of this sudden optimism stems from a calculation first performed by physicist Steven Weinberg with collaborators Paul Shapiro and Hugo Martel at the University of Texas, which in turn has its basis in one of the most slippery ideas in twenty-first-century physics, which is somewhat pompously called the "anthropic principle."

The anthropic principle is deceivingly simple to state and equally difficult to fully come to terms with. It is based on the suggestion that some, or perhaps all, of the fundamental constants in nature describing elementary particle interactions are what they are because if they took on different values, we wouldn't be here to measure them.

When one first hears this, it sounds like either a truism or a religious claim. But it is far from either. It does not imply, as some fundamentalists have tried to argue, that physics is on the verge of proving that the universe was created specifically for humankind to live in. Rather, at its best, it suggests that it is at least possible that, if the underlying theory of the universe does not uniquely predict the nature of particles and fields that can exist, then their may be no fundamental dynamical reason why the universe we live in is the way it is.

I should say at the outset that this idea goes completely against the grain of the entire history of physics over the past four hundred years. Generations of physicists have believed that their job was to explain why the universe had to behave the way it does, rather than why most possible universes would behave differently. Nevertheless, in the back of the minds of those physicists who have tried to derive new fundamental laws over the years, the nagging question asked in public by Einstein early in the past century has continued to burn a hole. As he put it, using a religious metaphor: "Did God have any choice in the creation of the Universe?"

By posing this question, Einstein in effect wondered whether there might be only one consistent set of laws that could result in a workable universe. Could it be that, if the electron was not 1/2000 of the mass of the proton, or if the electromagnetic force was not forty orders of magnitude stronger than gravity, the logical consistency of whatever underlying theory governs the physical workings of the universe would fall to pieces? Or, could one imagine a plethora of possible universes, each of which had different values for these quantities, and each of which could still form a logical and consistent whole? If the former is true, a Theory of Everything has teeth. If the latter is true, then physics is ultimately, as John Preskill at Caltech once put it to me an "environmental science," with even the fundamental laws of nature being determined by possible "environmental" accidents.

All of this metaphysical speculation began to take on greater significance in the latter part of the twentieth century as new ideas in physics spawned new ideas in cosmology. For example, once inflationary theory became widely accepted as a wonderful candidate idea to resolve various puzzles in the nature of observational cosmology, it was quickly recognized and stressed by the physicist Andrei Linde–one of the most inventive of the inflationary pioneers–that its principles would in general imply that the entire visible universe is likely to be merely a part of an incredibly complicated "metaverse" of causally disconnected universes. Some of these may be collapsing, others expanding, some may only now be experiencing a big bang expansion, and others may have long ago ended inside of cosmic black holes.

The possibility that many different universes might exist even in our mere three-dimensional space became compounded by the possibility that a higher ten- or eleven-dimensional space might settle into one of a virtually uncountable total number of possible ground states. The natural question then becomes: Did a single universe settle into a single ground state, or could it be that there are a host of different universes in a kind of "metaverse," each of which could settle into a different possible ground-state configuration?

For the most part, except for a cadre of philosophically minded theorists, no one much worried about this issue for a long time. Physicists are trained to calculate things from first principles, and moreover the remarkable successes of particle physics in the 1970s had demonstrated that it should be possible explain all of the observed phenomena at subatomic scales using three simple and elegant theories. It is true that in order for objects such as stars to operate and to be able to cook light elements such as hydrogen and helium into heavier elements such as carbon, oxygen, nitrogen, and all the other substances so vital to life on Earth, some remarkable coincidences seemed to be required at the level of nuclear physics. But, coincidences happen all the time, and indeed without knowing all of the spectra of possibilities, the fact that the observed route to stellar burning seemed to depend on some numerical fine-tuning was not particularly extraordinary, even if some intrepid speculators did write articles and books on the subject.

Then along came dark energy. Suddenly there was a parameter in nature that was so strange that no sensible explanation of its existence seemed within sight. Physicists began to explore possibilities that had otherwise seemed perhaps too distasteful, and Weinberg and his collaborators asked themselves the question: If there are possibly an infinite number of different universes, and if each universe could have a different value of the energy of empty space, what value might we expect to measure in a universe full of stars and galaxies that is over ten billion years old?

The somewhat surprising answer to this question is that one would expect, without knowing the details of what might be the a priori probability of having a universe with a given vacuum energy, that a universe in which galaxies could form after billions of years and astronomers could measure their properties would seem to require that this energy not be much larger than about five to ten times the vacuum energy we currently infer. Given that naive estimates based on quantum mechanics and relativity would suggest a value that is 10^{120} times larger, the anthropic argument provides an estimate that is far closer to the value we apparently live with in our universe.

At present, it is fair to say that this anthropic "explanation" of a vacuum energy that is comparable to the value we actually measure is one of the few viable proposals on the table. Having said that, however, it is important to realize that at this stage it is virtually impossible to know if this explanation really is an explanation at all. For example, while Weinberg and company did a calculation to show that if the vacuum energy alone were freely varying among all possible universes, one might expect a value comparable to what we see in our universe, without a fundamental theory that tells us which fundamental free parameters are variable, and which are fixed by fundamental laws, it is hard to know how seriously to take this simple first guess.

This, ultimately, is the fundamental problem in my mind with anthropic arguments. They may seem suggestive, but without a fundamental theory they can never be more than this. Indeed, as I have said on at least one public occasion, the anthropic principle is something that physicists play around with when they don't have any fundamental theory to work with, and they drop it like a hot potato if they find one.

Nevertheless, while my own biases about this notion are clear, it is fair to say that the moment one recognizes the possibility that multiple separated universes might exist, due either to separate inflationary phases in an otherwise infinite volume, or to the existence of higher dimensions, an anthropic explanation of fundamental parameters in our universe becomes at least a reasonable logical possibility. It is for this reason that a variety of sensible and distinguished individuals had begun to advocate this idea, and why it is at least worth examining further, even before string theory adds its two cents.

This finally brings us back to M-theory. Faced with the prospect that this theory may ultimately predict a virtually uncountable set of possible universes, some string theorists did a 180-degree about-face. Instead of heralding a unique Theory of Everything that could produce calculable predictions, they are now resorting to what even a decade ago they may have called the last refuge of scoundrels.

But, when string theorists take a position, they do it with flair. In attempting to graphically explore the different ground states of a subset of the set of all string vacua, some theorists realized that the diagrams looked like complicated landscapes, with billions and billions of sharp mountains and deep valleys. Physicists Joe Polchinski, Raphael Bousso, and Leonard Susskind felt that the images were so striking that they capitalized the description, and invented what they called "the landscape."

You can guess the argument by now. String theory/M-theory predicts more than 10^{100} possible configurations in which a three-dimensional universe might arise from a higher-dimensional framework (even though no one quite knows how many dimensions are truly fundamental). So, among all these vacua there are likely to be some with extremely small values for the vacuum energy, comparable in fact to what we measure today. These would be anything but generic universes, and would certainly not be what an otherwise unbiased observer would predict to find in a random universe. But, perhaps there are no unbiased observers! If observers like ourselves can exist only in universes that have at most an extremely small cosmological constant, then as long as the M-theory landscape provides that possibility somewhere, then that is where we will find ourselves.

What is perhaps most amazing about this is the degree to which this new reliance on postdiction is being adopted in parts of the community. In the end, it may be correct. It may be that string theory cannot predict from first principles a parameter as fundamental as the ground state energy of our universe. It may merely be an environmental accident, after all.

Still, this is a far cry from the excitement about a Theory of Everything raised twenty years ago during the first flush of enthusiasm associated with string theory, extra dimensions, and the new potential for unifying quantum mechanics and general relativity. Indeed, after the incredible journey of physics during the past century, after all the remarkable discoveries, theoretical and experimental, discussed in this book, this proposal seems rather like an anticlimax. As Edward Witten has commented, politely, about this approach: "I'd be happy if it is not right. I would be happy to have a more unique understanding of the universe."

His point is well taken. A cynical individual might suggest that some string theorists have embraced landscapes because since the theory cannot apparently predict anything anyway, it is gratifying to find a quantity that reinforces the notion that ultimately no fundamental constant in our universe is predictable. Nevertheless, as Witten's remark underscores, if the landscape turns out to be the main physical implication of the grand edifice of string theory or M-theory, then instead of precise predictions about why the observable universe of three large and expanding spatial dimensions must be the way it is, we might be left with the mere suggestion that anything goes. What was touted twenty years ago as a Theory of Everything would then instead have turned quite literally into a Theory of Nothing.

But the good news is that we don't yet know. The more we explore the ideas of string theory, M-theory, and Braneworlds, the more it becomes clear that we understand far less than we thought about what might be possible in nature. Even the fundamental concepts of strings and dimensions–which lay at the heart of the original 1984 revolution–may now be beginning to melt away.

Will whatever physical theory results in the aftermath of all this, following whatever discoveries are made by experimentalists in the coming decades and by theorists in the coming centuries, resemble any of the speculative, if beautiful, mathematical notions at the heart of the current focus of research? That, I believe, is anyone's guess. I have recently discussed this question with two active string theorists, John Schwarz and Nati Seiberg, and perhaps not surprisingly both still feel that the mathematical insights already gleaned from string theory are so powerful that whatever ultimate theory we may derive for the workings of nature at fundamental scales, it will contain at least the germ of present string theory ideas.

I admit that, during the course of thinking about these issues as I have written this book, I myself have run hot and cold. There have been moments when the remarkable depth of the mathematical insights being explored in the course of recent years has left me awed, and there have been times when the sheer hubris of the claims, and the lack of associated results has left me shaking my head in disbelief.

But I want to make it clear that while I think it is certainly possible and, given historical perspective, perhaps even likely that all of the formalism currently being explored is a mere house of cards, and that it might tumble as soon as the force of some new experiment or observation overwhelms it, this does not mean the effort is not worthwhile.

If the joy of the search exceeds the pleasure of the finding, then we continue to be joyfully engaged an intellectual struggle that shows no signs of ending and in which hidden universes have always been a part. To make progress in our attempt to understand the universe at its most fundamental level, we need to fearlessly open up new paths into otherwise unexplored places, and we must not be afraid of wrong turns and dead ends, even if, like the ether squirts of the nineteenth century, that is what ideas such as grand unification or string theory ultimately turn out to be.

No doubt we are hardwired to believe that the universe of our experience cannot be all that there is. This would certainly explain the persistence of religious faith in an apparently unfair world of toil and struggle without obvious purpose. Perhaps that, too, is why we keep returning to the notion that just beyond our reach, just behind the mirror, lies the key to knowledge.

But even if in the end this longstanding pursuit of extra dimensions proves to have been a grand illusion, generations of dreamers have been inspired by it to keep on dreaming, and generations of seekers to keep on seeking. We have learned and will in the process continue to learn more about nature and our own place within the cosmos. And I believe one could make a good argument that such efforts make life worthwhile.

For those who may be less romantic, there is another plus. In our continued and possibly flawed search for hidden universes and extra dimensions, we are certain to stumble upon unexpected and undoubtedly unrelated natural wonders that are currently beyond our wildest imagination, and that may have a direct impact upon our own future. If the past is any guide, one thing seems certain: The universe always seems to come up with new ways of surprising us.

EPILOGUE
TRUTH AND BEAUTY

In . . . Philosophical Theories as well as in persons, success discloses faults and infirmities which failure might have concealed from observation.

–John Stuart Mill, *On Liberty*

On January 30, 1991, the physicist John Bardeen died. An obituary appeared in various major papers around the country, but most people then, like most people now, would hardly recognize the name–in spite of the fact that it is arguable that Bardeen changed the face of the twentieth century as much as any other scientist of his era. He was the only physicist ever to win two Nobel Prizes in physics. The first was for the invention of the transistor, which, as I have mentioned already, is at the very basis of almost all of modern technology. The second was for the explanation of superconductivity, the remarkable property of some materials to allow currents to flow without resistance of any kind below a certain temperature, a phenomenon whose technological impact will most surely grow in this century.

Yet, even among lay people with an interest in science, I would venture to suggest that there is more interest in string theory than superconductivity, in spite of the fact that the former has yet to have any clear impact on our understanding of the physical universe, much less our daily lives.

This is not meant to be judgmental. Rather, it simply reflects something that I think is deeply ingrained in the human psyche. "Space" and "time" are among the very first concepts that are framed as our own consciousness emerges shortly after the fog of birth. So it is not surprising that considerations of the ultimate nature of space and time may continue to appear more interesting than the things that merely happen within space and time.

I began this book wondering about what drove an ancient ancestor to leave an imprint of his or her child's hand on a cave wall. I suggested that it was to create a measure of permanence, something that might live on, as it in fact did, long after the participants in this artistic enterprise were gone. Time is our ultimate enemy, and to conquer time means first trying to understand it.

Time is a subtler concept than one might imagine. Both future and past are not directly experienced, but must be intellectualized. Space, on the other hand, while immediate and visceral, nevertheless taunts us with its mysteries every time we do something as simple as gazing out at the horizon. Recall that for early European sailors the horizon represented the end of a world that we now know has no end. If we can be so easily fooled here on Earth, what do the more exotic mysteries that lie out in the darkness of the night sky hold for us?

Yet recall that I also ended the first chapter of this book with a warning from the famous French chemist Antoine Lavoisier about guarding against flights of the imagination regarding things one can neither see nor feel. His warning, of course, continues to go unheeded. Indeed, this book pays homage to the history of the remarkably constant human impetus, both scientific and artistic, to first imagine and then explore the reality that exists beyond our direct sensory experience.

Nevertheless, in spite of all the excitement regarding the possible existence of extra dimensions, I confess yet again to being an agnostic. Perhaps it is more appropriate to call myself a skeptic. This position sometimes gets me into trouble, especially in public debates, but I am nevertheless proud to be part of a noble tradition in science. I earlier referred to Richard

Feynman's statement that science is "imagination in a strait-jacket." Most good ideas are wrong, in that nature does not choose to exploit them. If that were not the case, doing science would be far easier.

I do remain fascinated with the myriad possibilities for new and hidden realities afforded by extra dimensions, but I try to temper my enthusiasm with the realization that, like Fox Mulder, I "want to believe." Large, hidden extra dimensions are seductive, and I wish that they were true in the same sense that I wish I could use a warp drive to travel to distant stars, to go where no man or woman has gone before. We may indeed be on the threshold of discoveries that will truly change everything, that will further inspire a generation of artists and writers, and vindicate once again the wildest imaginings of science fiction writers. But there is no evidence at this time that any such imminent breakthrough is likely or inevitable. There are beautiful theoretical arguments that are strongly seductive, as I have tried to describe, but there were beautiful theoretical arguments in 1970 that were also strongly suggestive–but also wrong–that string theory might provide a fundamental theory of the strong interaction.

Equally beautiful theoretical arguments prompted Kaluza and Klein to make their bold proposals, but we now understand those elegant concepts were introduced before their proper time. Kaluza and Klein could never have known that the theory they were exploring was missing key features of reality, including two of the strongest forces in nature. Perhaps we are in the same boat today.

Today's confused and tentative explorations of possibly infinite extra dimensions and infinite landscapes of extra-dimensional worlds must be seen as simply the most recent expression of a longstanding scientific and cultural tradition. One can marvel, for example, at the remarkable resemblance between the claim that elementary charges in our space are merely the ends of fundamental strings that may stretch out into higher dimensions, and the nineteenth-century claim that these charges were "ether squirts"–places where a four-dimensional ether flowed into our three-dimensional world.

Such eerie resemblances imply neither that current science is pure fiction, nor that the ill-founded speculations of the 1870s bore some hidden truth. To make such arguments would be just as misplaced a notion as subscribing to the claims that a resemblance between ancient Eastern mystical writings and some of the tenets of quantum mechanics implies the ancient writers had any idea of even what hydrogen was, much less how to calculate the spectrum of light emitted by it.

Similarly, it has been stated many times since 1984 that the remarkable discovery of string theory in the 1970s and its rediscovery in the 1980s was a unique situation in the history of physics: We were living in the twentieth century, having accidentally discovered the physics of the twenty-first or twenty-second century. That could, in fact, be true. But we have no proof that it is or was. It is just as likely to be true that we are instead reliving the delusional enthusiasm for the extra dimensions of the nineteenth century. That is also cause for neither despair nor hope, in my opinion. It is simply an inevitable product of living in confusing times. But being confused *is* cause for hope. Perhaps there is no state more desired by theoretical physicists than being confused, for it is confusion that compels us to seek out new knowledge and the opportunities for breakthroughs.

As we thus celebrate the remarkable ideas that have emerged from the solid scientific progress of the past century, we must be careful to keep things in perspective. I can think of no better way to do this than to relate the intertwined discussions of three of the most accomplished theoretical physicists of my own generation: David Gross, Frank Wilczek, and Edward Witten. All three have played important roles in the stories related in the preceding pages.

When I first told Wilczek that I was writing this book, he related a somewhat disconcerting story to me about a time when he tried to explain the remarkable aspects of the strong interaction between quarks to a public audience (before the Nobel committee anointed this work as being important). After the talk, a member of the audience raised his hand and asked: "Why should I care about all of this? Isn't it just the four-dimensional manifestation of the far more fundamental predictions made by string theory in ten dimensions?"

This reminded both of us of an earlier time when we were working together to advise the Smithsonian Institution on several projects it was sponsoring, supported by the Defense Advanced Research Projects Agency (DARPA, a national security funding group), on the detection of neutrinos.

DARPA was interested in detecting neutrinos because they are emitted by nuclear reactors, and nuclear reactors are on submarines, and detecting submarines is of vital strategic importance. Thus, even far-out schemes seemed to DARPA to be worth throwing a bit of money at, because if any of them worked, it could easily have tipped the Cold War strategic balance in our favor.

Of the projects we examined, all were rather fanciful, but one was at least marginally plausible. It was a proposal to detect neutrinos from possible nearby nuclear weapons tests using a large ton-sized detector. However, when we informed DARPA of our choice, we were told that they had already been supporting the work of a well-known (but misguided) scientist, who claimed he had a bread box–sized device that could detect neutrinos from every nuclear reactor and nuclear weapon on Earth. How could DARPA therefore justify funding a ton-sized detector near a nuclear weapons test when it was spending millions on a far smaller detector that was argued to be far more sensitive?

This is the problem that often arises when speculative science is valued more than the remarkable achievements of empirically tested science. The moral for our present discussions is, I hope, clear. The tremendous intellectual efforts over the past century to formulate a candidate theory that might unify quantum mechanics and gravity in a higher-dimensional framework should not be minimized. The theoretical and mathematical results that have been developed are fascinating. But neither should they be celebrated for more than they yet are.

It does a disservice to the most remarkable century in the history of human intellectual investigation to diminish the profound theoretical and experimental discoveries we have made in favor of what is at the present time essentially well-motivated, educated speculation. It is also simply disingenuous to claim that there is any definitive evidence that any of the ideas associated with string theory yet bear a clear connection to reality, or that they will even survive in their present form for very much longer. Perhaps more to the point, the deeper we probe these theories, the hazier they seem to have become.

Which brings us to Edward Witten, who has been the leading force driving string theory since the mid 1980s. Ed is not only an incredible intellect, but he is also a refreshingly honest one. He says what he means, and he always has a sound reason for saying what he does.

Edward is also the attributed author of the infamous statement regarding twenty-first-century physics in the twentieth century, which is probably one reason it is so often repeated. But one should not read more into that observation than I believe Ed intended. Ed may be a "true believer" in string theory, but that simply reflects the very nature of his position on the theoretical forefront. It is, as I have stressed, very difficult to devote the incredible intellectual energy and focus that are required over long periods of time in the attempt to unravel the hidden realities of nature if one does not have great personal conviction that one has a good chance of being on the right track. As Edward said succinctly at a recent meeting on the future of physics, regarding why one should study string theory: "I don't consider it plausible that a completely wrong theory would generate so many good ideas."

The same level of personal conviction is required of artists and writers, as well. But what makes science somewhat different, I believe, is that great scientists are prepared to follow an idea for as long as decades, but at the same time are equally prepared to dispense with all of this effort in a New York minute if a better idea or a contradictory experimental result comes along.

With this in mind, a number of other statements that Edward made at this recent meeting are quite telling and, I believe, validate the gestalt I have tried to characterize here. Summarizing the essential progress of the theory he has devoted much of the past two decades to studying, he said: "It [string theory] is a remarkably simple way of getting a rough draft of particle physics unified with gravity. There are, however, uncomfortably many ways to reach such a rough draft, and it is frustratingly difficult to get a second draft." He next reiterated that while we lack any understanding of the core idea–equivalent to the Equivalence Principle (between gravity and acceleration) that was at the heart of general relativity–behind string theory, at its heart is the notion that space-time is an "emergent" and not a fundamental concept. Thus, the whole notion of what an extra spatial dimension may mean within the context of string theory is not clear. More interesting still, he argued that even strings themselves are not likely to be fundamental, but that they, too, would prove to be an emergent concept based on something more fundamental.

Finally, Witten stressed what I believe, given the current situation in string theory after more than twenty years of research, is an eminently reasonable position: That it is at best plausible that we will manage to ever understand what string theory is all about, and, whether or not we do, that it is not at all clear whether we will be able to use it to understand nature. This will depend upon factors beyond our control, including how complex the ultimate answer may be, and what clues we might be lucky enough to derive from experiment. I reiterate that these were statements made not by a skeptic, but by someone who passionately believes that string theory contains a germ of truth.

I have debated this very point on stage twice with Brian Greene, who has worked as hard as anyone to popularize and celebrate string theory. Brian earnestly and successfully communicates the excitement of the theory in a way that can inspire lay people. Brian is an honest popularizer and prefaces his remarks about string theory with qualifications about its present speculative and unproven state. However, he is so convincing and enthusiastic that I have argued with him that when such things as animations of strings within elementary particles are presented, even if the intent is merely illustrative, it tends to give the impression that string theory is a better defined construct than it currently is, and also suggests it gives definite predictions about the properties of observed elementary particles in four dimensions. This is of course a subjective issue, and I know Brian disagrees with me about this. Ultimately, from my perspective, this enthusiasm is unwarranted at the present time, given what might be described as the current impotence of the theory.

In order to dramatize my own concerns about the dangers of conveying enthusiasm as truth, I have claimed that string theory is in some sense the least successful great idea in twentieth-century physics, a statement that *The New York Times* kindly quoted out of context. At a recent event celebrating Einstein, I pointed out that it is somewhat incongruous to, on one hand, portray as tragic the past thirty years of Einstein's life, during which he worked on his own on an unsuccessful unified field theory, while at the same time celebrate at scientific meetings and in the popular media perhaps three thousand man-years of full-time intellectual activity by a brigade of some of the most talented young theoretical minds around the world on a proposed unified theory that has thus for been largely fruitless in its predictions, and has yet to be properly understood.

I believe both extreme viewpoints are inappropriate. Einstein's efforts were no more tragic than the recent string program has been an unqualified success. Both are part of the search for underlying order in the natural world that proceeds by fits and starts, and is full of far more blind alleys than awakenings.

And I want to reiterate once again, perhaps even more strongly, that these efforts, even if they do not produce the results we wish, will not have been wasted. Ed Witten wrote me a frank letter after I asked him to read a draft of this book. He described how, when he was a student in the 1970s, he was obsessed with trying to understand, on the basis of simple analytical physical calculations, exactly why quarks are confined together. He gave up, because he thought to the problem was too hard. Now, almost thirty years later, he is working on the problem again, this time using the tools of string theory, and he feels he is making progress. As he put it: "Being able to develop these models in the last decade, fifteen years after giving up on quark confinement as too hard, has been a lot of fun." Moreover, after arguing that the many developments I have discussed are evidence that our understanding of string theory is reaching a deeper level, he nevertheless emphasized that this most recent work, on using string methods to attack quark confinement and not quantum gravity, as originally intended, has "maybe been the most fun for me."

One never knows where insights will come from, or where they may lead. The pleasure of research is discovering the unexpected. Ed's poignant remark underscores that ultimately the driving force behind all human inquiry is the satisfaction of the quest itself. We may or may not be hardwired to long for hidden realities, but we are most certainly hardwired to enjoy solving puzzles, especially when their resolution is far from what one may initially have expected.

I would also be less than candid if I did not reveal that there is other, more personal evidence I now have that the string effort has already borne some fruit. After *The New York Times* published my supposed statement on the failures of string theory, I received a package in the mail from California. Upon opening it, I found a fruit basket from John Schwarz with a note, which read: "Dear Lawrence: Now maybe you won't feel it's all been so fruitless."

This finally brings me to David Gross, who has played the most interesting sociological role in the story I have told. You will recall that David was a student at Berkeley in the 1960s, the era of bootstrap models and dual string models as applied to strongly interacting elementary particles. He thus received his scientific grounding in theories that turned out to be footnotes in scientific history. But it was ultimately his own work on asymptotic freedom, for which he has shared the Nobel Prize, that turned them into footnotes.

In another poetic example of the ironies of scientific progress, over a decade later David became a key part of the new string revolution, which reinstated the very ideas he had earlier killed, but this time in a new context. His work on heterotic strings, and the possibility of explaining all the phenomenology of elementary particle physics in four dimensions via an underlying theory in ten *and* twenty-six dimensions, helped to create the fervor that motivated Witten's statement about the incursion of twenty-first-century physics into the twentieth century.

But, as Ed Witten has admitted, these ideas ultimately just produced a "rough draft" that has yet to ever go beyond this stage. One might think that having witnessed the demise of a similar rough draft in the 1970s that Gross might temper his statements about the ultimate truth of the new string theory. But it is an interesting facet of the human condition that revolutionaries sometimes replicate certain features of the regimes they set out to overthrow. In this case the former young rebel has become something of a defender of the faith.

In many forums David has argued forcefully–and, of course, brilliantly, because his is a powerful intellect–that the theory is simply too beautiful *not* to be true. As such, every new result tends to merely reinforce its truth, even without the luxury of experiment. As I have described, this attitude has been adopted by many of the younger researchers in this field, who are, of course, strongly influenced by their senior mentors, as well as the mathematical appeal of the subject.

I do not mean to cast aspersion on David's scientific work, which has been impeccable and important. And as I said, it is perfectly reasonable to expect those theorists who have devoted decades to exploring a theory to be driven by an expectation of its inherent validity. The problem, however, is that this viewpoint strikes some, including me, as sounding like religion more than science.

At other times in this century, science may have been able to more easily tolerate such confusion. Perhaps I am oversensitive on this subject, but I have spent much of the past several years fighting attacks on science, from the classroom to the White House. The aim of both these sets of challenges has been to replace the hard-won results of the scientific process with ideological dogma. In the former case, where individuals have been attempting to impugn a well-tested scientific theory that is the foundation of all of modern biology, I have often been told that science itself is merely another kind of religion. I believe that nothing should be further from the truth, and anything that confuses this issue is regressive.

Still, the convergence of truth and beauty, at least as we behold it, is a notion that is in some sense central to almost everything I have discussed in this book. Indeed, I began with a discussion of the mysterious fact that nature and beautiful mathematics seem inextricably united. Recall Bertrand Russell's description of mathematics as possessing "not only truth, but supreme beauty." With that in mind it is, I believe, generally appropriate to give the last word to a mathematician whose work played a central role in the earliest developments that I have described here.

I refer to Hermann Weyl, the brilliant mathematical physicist whose results originally inspired Kaluza to ponder extra dimensions, and who first exposed the fundamental symmetry of nature that we now call gauge invariance, which is at the heart of the description of all the known forces in nature, including gravity. Weyl was a student of Hilbert, one of the fathers of higher-dimensional geometry, and, as you may also recall, a competitor of Einstein's on the road to developing general relativity. And Weyl ended his career at the Institute for Advanced Study at Princeton along with Einstein. So it is, in fact, particularly appropriate to turn to Weyl for enlightenment as we reach the end of our own journey through the looking glass.

Upon reflecting upon his work, which clearly touched not only on mathematics but on the physical world, Weyl made a profoundly insightful confession that appeared in his own obituary, written by the physicist Freeman Dyson in 1956. Nothing I can think of better captures the dilemma exemplified by our ongoing, and remarkably timely, love affair with extra dimensions. Referring to his research, Weyl admitted:

> My work always tried to unite the true with the beautiful, but when I had to choose one or the other, I usually chose the beautiful.

So it is that mathematicians, poets, writers, and artists almost always choose beauty over truth. Scientists, alas, do not have this luxury, and can only hope that we do not have to make a choice.

ACKNOWLEDGMENTS

Each of my books has been a tremendous learning experience. I depend greatly both on previous authors with their accumulated wisdom and on generous colleagues from a variety of areas who help steer me in the right direction as I begin to grapple with sometimes totally new subjects.

Thinking about the cultural and social legacy associated with the notion of extra dimensions, I am enormously grateful to my Case colleague Henry Adams, professor of art history and former curator at the Cleveland Museum of Art. He and a student of his provided me with a wonderful bibliography of twentieth-century artists whose work related to the notion of a fourth dimension.

One of the most important books that I initially turned to was Linda Dalrymple Henderson's *The Fourth Dimension and Non-Euclidean Geometry in Modern Art*, a wonderfully broad and complex discussion of the notion of a fourth dimension in modern art. I owe a great debt to her scholarship, which pointed me in the direction of numerous important sources, and which has inevitably strongly influenced my own thinking.

I also owe thanks to several individuals in the science fiction community, notably Charles Brown and several other of my colleagues on the board of the Science Fiction Experience in Seattle, who helped direct me to the appropriate literature, in particular to the fiction and nonfiction work of Rudy Rucker on the fourth dimension.

I want to thank the librarians at the Case Western Reserve University for providing me with great assistance, including a room to work and store books in, as I tried to devour the appropriate literature and write in a quiet place away from my office.

I thank numerous colleagues for discussions and illuminations related to the physics and historical ideas discussed here. I learned a great deal from the informative introduction in the book *Modern Kaluza Klein Theory,* by Tom Appelquist, Alan Chodos, and Peter Freund, as well as the comprehensive two-volume opus *Superstring Theory* by Michael Green, John Schwarz, and Edward Witten, and the more recent two-volume work *String Theory* by Joe Polchinski, as well as, of course, the numerous papers in the literature so easily accessible thanks to the physics Web archive created by Paul Ginsparg.

I owe a personal great debt to my Case colleague and friend Cyrus Taylor, who spent many hours introducing me to the intricacies of string theory, and who directed me to appropriate places in the literature. Gia Dvali helped me first tackle large extra dimensions, and continues to provoke my imagination. My friend and colleague Frank Wilczek has shared many of his physics and philosophical insights with me over the years, and I appreciate his willingness to provide feedback on some of the issues I raised in the final chapters. Lastly, I want to thank all of my colleagues who have looked at this manuscript and provided comments and feedback. In particular, I want to thank Glenn Starkman for general suggestions, and John Schwarz and Edward Witten for carefully reading the sections on string theory and making comments.

Finally, I want to thank once again my wife Kate and daughter Lilli for putting up with and supporting me through yet another book project, and for being joyful reminders that physics is just one part of a fascinating world.

GLOSSARY

Alpha rays: Rays made up of the nuclei of helium, containing two neutrons and two protons, which are produced in the radioactive decays of various heavy nuclei.

Angular momentum: A twisting force imparts angular momentum to objects, causing them to spin. Angular momentum is calculated as the product of the mass of an object times its rotational speed.

Anomaly: Due to quantum mechanical effects, a symmetry of nature that appears in a classical theory (such as electromagnetism) can be violated at the quantum level. When this happens the symmetry is said to be anomalous, and the quantum mechanical contribution that violates the symmetry is said to be an anomaly. Several "anomalous symmetries" are known to exist in nature. However, it is very important that quantum mechanical effects do not spoil the gauge and general covariance symmetries that are at the heart of the four known forces in nature. Making sure this does not happen has played a key role in efforts to develop string theories as candidate theories for the natural world.

Antiparticles: The laws of quantum mechanics and special relativity together imply that every elementary particle in nature must have an antiparticle, with equal mass and opposite electric charge. Many antiparticles have been created in the laboratory, and are used regularly in high-energy particle accelerators that explore the nature of matter and energy at fundamental scales. When particles and antiparticles collide, they can annihilate, producing pure radiation. Some neutral particles can be their own antiparticles.

Asymptotic freedom: The remarkable property of the strong interaction, discovered in 1974, that the force between quarks becomes stronger as you pull the quarks apart. This is the opposite behavior from electromagnetism, which gets weaker as elementary charges are moved far apart from each other. Asymptotic freedom is presumably related to the fact that no observable isolated quarks exist in nature (a phenomenon called confinement).

Beta rays: Rays made up of electrons, which are produced in the radioactive decays of various elementary particles and nuclei.

Black body radiation: When a perfectly black solid, like the heating element on a stove, is heated up, it emits a continuous set of colors of radiation, changing from red hot to blue hot to white hot, for example. This distribution of radiation uniquely determines the temperature of the object, and was explained using the laws of quantum mechanics early in the twentieth century.

Bootstrap model: An idea that achieved prominence in the 1960s in response to the growing number of strongly interacting elementary particles, which suggested that no elementary particles were truly fundamental, but rather that all particles could be made up of other elementary particles. It proposed instead that what *was* fundamental was the mathematical relationship between particles that governed their interactions with each other. Bootstrap models eventually led to the development of string theories that attempted to describe the interactions of strongly interacting elementary particles.

Bosons: Elementary particles in which the spin angular momentum is quantized, having a value equal to an integer multiple of some fundamental quantum of angular momentum.

Chirality: Certain objects, like our two hands, or the spiral structures that make up DNA can be said to be left handed or right handed, i.e., mirror images of each other. This property is called chirality. Elementary particles with spin angular momentum can be chiral, in that they can appear to be spinning in either a clockwise or a counterclockwise direction about their spin axis. One type of particle is called left handed, and the other right handed. Theories that distinguish between left- and right-handed particles are called chiral theories. The weak interaction is one such example as, for example, only left-handed neutrinos appear to sense the weak interaction. (As a result we do not even know if right-handed neutrinos exist in nature.)

Cloud chamber: A device developed in the early part of the twentieth century that produces observable tracks when charged elementary particles, such as the particles in cosmic rays, traverse the chamber. When these particles more through the chamber the gas vapor surrounding the particles with which they collide condenses, producing a visible vapor trail. Different particles produce qualitatively different tracks.

Compactification: In theories with extra dimensions beyond the three space and one time dimension of our experience, one has to explain why the other dimensions are not observed. One solution involves compactification, in which the extra dimensions are curled up into "balls" that are so small that no experiment yet performed could detect their existence. The process by which one goes from a higher-dimensional theory to an effective four-dimensional theory is called compactification, and trying to understand how this might occur is one of the major challenges of string theory, and other higher-dimensional theories.

Conformal invariance: A mathematical symmetry that encompasses not only the general covariance that is at the basis of general relativity but extends it to include so-called scale transformations. If the world were conformally invariant, then the world would appear unchanged if I doubled the size of all objects, their masses, etc. This is clearly not the case, so conformal invariance is not a property of the real world as we measure it. However, it is an underlying property of string theories, and clearly one of the challenges of having string theory touch base with the world that we observe is to find mechanisms by which this symmetry is broken in our world.

Connection tensor: A mathematical quantity that encodes the geometric nature of space. The connection tensor in particular explains how the length and orientation of a standard ruler might be measured to change as it moved between nearby points in a curved space. The connection tensor therefore encodes information about the curvature of space.

Cosmic microwave background: The afterglow of the big bang, this radiation is a remnant from the earliest era of the expansion, when the temperature was so high that matter and radiation were in thermal equilibrium. Once the temperature had cooled sufficiently (to about three thousand degrees above absolute zero), protons and electrons began to be able to combine to form neutral atoms, which decoupled from the radiation so that the universe became transparent. The remnant radiation cooled as the universe expanded, and is now at a temperature of about three degrees above absolute zero.

Cosmic rays: Energetic elementary particles of many different types that bombard the earth regularly from space. They originate from locations as close as our own sun, and as far away as the centers of distant galaxies.

Dark Energy: When we add up the total amount of mass in the visible universe, and compare it to the total energy needed to result in the flat universe (see *Flat universe*) that we appear to live in, there is a factor of three too little mass to account for the flatness of space on large scales. At the same time, the observed expansion of the universe appears to be accelerating, which could only be the case if empty space possessed energy (see *Vacuum energy*). The amount of energy needed to result in the observed acceleration turns out to be precisely that required to also account for a flat universe. We currently understand very little about this "dark energy," which resides in empty space, and do not know if it is vacuum energy, or some other kind of yet more exotic form of energy.

D-branes: Multidimensional surfaces (generalizations of two-dimensional membranes–hence the name) on which "open strings" that is, strings that are not closed loops, and that propagate in higher dimensions, can end. The "D" in D-branes does not refer to the dimensionality of the brane, but rather to the specific boundary conditions that are imposed at the end of the string as it merges with the brane. D-branes are now understood to be very important objects within string theory, though they were not known in the earliest formulations of the theory.

Density fluctuations: Observed stars, galaxies, planets (and ultimately people) initially arose as very small inhomogeneities in the distribution of matter and radiation in the early universe collapsed due to their internal gravitational attraction. Regions where there was a very small excess of matter, for example, compared to the background value, would expand slightly more slowly than the background, eventually becoming so much more dense than the background that they decoupled from the expansion of the universe, and started to collapse. We believe this is how all large-scale structures now observable in the universe first formed. The question then becomes, what caused these initial density fluctuations in the early universe? We currently have reason to believe that they formed due to the quantum mechanical effects at very early times, as a result of inflation.

Electron: An elementary particle with negative electric charge that comprises the outer parts of all atoms. Neutral atoms contain an equal number of electrons and protons, with the latter existing within a dense nucleus at the center of the atoms. As far as we know, the electron is absolutely stable.

Equivalence principle: The principle that all objects fall at the same rate in a gravitational field. Einstein argued that this is equivalent to the notion that in a local free-falling frame, the effects of gravity will be unobservable. This principle formed a fundamental pillar of his general theory of relativity, because it allowed him to present a completely geometric description of gravity in which its effects could be ascribed to the curvature of space.

Ether (also *Aether*)*:* The hypothetical substance that was believed for centuries to fill space and in which it was believed that light waves propagated.

In 1887 the physicist Albert A. Michelson and his colleague, chemist Edward Morley, demonstrated experimentally that the ether, as a medium in which light traveled, did not exist. Later, in 1905, Einstein demonstrated that the existence of such an ether was in fact inconsistent with the laws of physics.

Event horizon: A region surrounding a black hole, from which classically nothing, even light, can escape. As a result, once objects cross the event horizon observers outside of the black hole lose all information as to their future behavior.

False vacuum: If we describe the vacuum state as the lowest energy state in which a system can exist (such as a region of empty space devoid of matter or energy), a false vacuum occurs when the lowest energy state in certain circumstances turns out not to remain the lowest energy state as those circumstance change. Possible examples include when the value of some external field, or the temperature of the system, changes. The system may exist in this false vacuum state state for a long time, but it will eventually decay, by the rules of quantum mechanics, into the new lower energy state, releasing energy in the process.

Fermions: Elementary particles in which the spin angular momentum is quantized, having a value equal to a half-integer multiple of some fundamental quantum of angular momentum.

Flat universe: General relativity implies that space can curve in the presence of mass and energy. On the largest scales, if light travels in straight lines, this implies that the universe is spatially flat. A spatially flat universe is infinite in extent, and, if dominated by matter, will continue to expand forever, with the expansion rate slowing asymptotically, but never quite falling to zero. We appear to live in a flat universe, as far as we can tell, although not one dominated by matter.

Gamma rays: The most energetic electromagnetic rays. The photons making up gamma rays can have energies as great as or greater than the energy associated with the rest mass of elementary particles such as electrons and protons.

General covariance: A mathematical notion at the heart of Einstein's general relativity theory that implies that the laws of physics are independent of any specific coordinate frame in which we choose to measure them. One of the implications of this is that for an observer in free fall in a gravitational field, the effects of gravity will appear to disappear. Another is that an observer accelerating upward in an elevator in empty space will experience a force pushing him toward the floor that will be completely indistinguishable from the force of gravity that he would experience if he was at rest in a gravitational field.

Grand unification: The theoretical notion that the three nongravitational forces in nature–the weak, electromagnetic, and strong forces–can actually be unified in a single framework, and moreover, that at a very small scale, perhaps fifteen orders of magnitude smaller than we can measure today, all of these forces will appear to have the same strength.

Grassmann variable: A mathematical quantity that has some properties of a normal number, but nevertheless has some vastly different properties. For example, when a Grassman number is multiplied by itself, it produces zero. Two different Grassman variables, A and B, when multiplied together in one order, say AB, equal the negative value when multiplied in the other order, so that $AB=-BA$. It turns out that these properties mimic the quantum mechanical properties that govern the behavior of fermions.

Graviton: When one combines quantum mechanics and relativity, all forces are conveyed by the exchange of elementary particles, like the photon, the fundamental quantum of electromagnetism. We call the hypothetical particle that conveys gravitation the graviton. Individual gravitons have not yet been measured because of the weakness of gravity, although we have no reason not to believe they exist.

GSO construction: A particular construction in string theory in ten dimensions, associated with the names Gliozzi, Scherk, and Olive, which removed the unwanted tachyon modes by introducing supersymmetry on the strings.

Hadrons: Elementary particles that have strong interactions with other particles.

Heterotic string: A string theory involving closed string loops in ten dimensions in which the different excitations of the string, moving in different directions along the string, behave quite differently. In fact, the left movers appear to live in a different number of dimensions than the right movers. In this way, it turns out that one can have consistent string theories in ten dimensions instead of twenty-six dimensions. Moreover the gauge symmetries that one hopes might be associated with the observed gauge symmetries in our world arise naturally as a part of this construction.

Hierarchy problem: Gravity is much weaker than all of the other forces in nature. This extreme hierarchy of forces is currently not understood, and is one form of what is often called the hierarchy problem. Another example is that the length scale at which the strength of all the nongravitional forces appears to become the same–the length scale at which grand unification is thought to occur–appears to be very much smaller than the scale associated with the size of particles such as protons and neutrons, and nuclei. It turns out to be very difficult mathematically to devise theories in which this is the case, and trying to resolve this difficulty is the hierarchy problem.

Hubble constant: In a uniformly expanding universe the recession velocity of distant objects away from us is proportional to their distance from us. The quantity determining the precise numerical relationship between velocity and distance is named the Hubble constant, in honor of Edwin Hubble, who first discovered this relation. Note that this quantity is not in fact a constant over cosmological times for most cosmological models.

Hypercube: Another name for a four-dimensional cube (tesseract).

Inflation: The idea, based on notions coming from the physics of elementary particles, that at very early times the universe underwent a brief period of extremely rapid expansion, during which distances increased by a factor greater than a billion, billion, billion, billion, in a fraction of a second. Such an expansion can naturally occur as the universe expanded and cooled at early times if there was a phase transition associated with a grand unified theory (see *Grand unification*), and can moreover explain all of the observed features of the universe today on the largest scales we can measure.

Large hadron collider (LHC): The new large proton-proton collider being built at the European Center for Nuclear Research (CERN) in Geneva. Planned to come online by 2007–2008, it will achieve energies large enough to explore for the mechanism underlying the origin of mass of elementary particles, and may reveal other new phenomena such as supersymmetry and possible large extra dimensions.

Local supersymmetry: This involves the mathematical formalism in which gravity and supersymmetry are combined together in one framework. One consequence of this is that the graviton, the fundamental quantum thought to convey the gravitational force, must have a fermionic partner, called the gravitino.

Matrices: Mathematical objects which take the form of tables of numbers with separate entries in the different rows and columns. Matrices can be multiplied together, added together, etc. and have thus have their own kind of algebra that is more complex than the algebra of simple real numbers. One of the eleven-dimensional limits of string theories that form a part of M-theory involves a description of nature in which matrices form the fundamental quantities akin to the numbers that describe positions in our four-dimensional space.

Metric: The mathematical quantity, called a tensor, that determines how physical lengths are measured in terms of the coordinates one uses to label the points in some space. For example, on a sphere, the physical distance between neighboring lines of longitude decreases as one moves to the poles. The metric tensor contains this information of how the distance between lines of longitude changes as you move around the surface of the sphere.

Moduli fields: In extra-dimensional theories such as string theory there are usually dynamical "fields" observable in our three-dimensional world that are associated with the actual radius of the presumably compactified and unobservable extra dimensions. These fields are called moduli fields, and their dynamics can either cause interesting new effects that might be measurable in our space, or cause severe empirical problems for model builders.

Momentum: A force acting on an object over some time imparts momentum to that object. For objects moving slowly compared to the speed of light, the momentum of the object is given by multiplying the mass of the object by its speed.

M-theory: The eleven-dimensional theory that is thought to underlie all known ten-dimensional string theories. Its existence was suggested once it was recognized that D-branes must be included in string descriptions, and these clarified the relationship between formerly disparate string models, suggesting some evidence of a yet higher dimensional theory. To date, no one has a clear understanding of the precise nature of this theory, or even what its fundamental variables are.

Muon: An unstable elementary particle, with a lifetime of one millionth of a second, that appears to be identical to the electron, except that its mass is about two hundred times greater. When it was first observed, the physicist I. I. Rabi uttered, "Who ordered that?"

Naturalness: In physics formulas one often finds numbers comparable to unity, such as 2 or pi. However, physicists call it "unnatural" when one finds in a formula a very large or very small dimensionless number, like 0.0000000000000000000000001 or 35,000,000,000,000,000,000,000,000,000. The ratio between the strength of gravity and electromagnetism is one such very small number, for example, which is why the hierarchy problem is one form of a naturalness problem.

Neutrino: A light neutral particle produced in the radioactive decay of a neutron (and various other particles). The neutrino has no electromagnetic or strong interactions, and thus interacts so weakly with matter that neutrinos produced in the decay of a neutron can, on average, travel right through the Earth without a single collision or interaction.

Neutron: A neutral elementary particle with a mass comparable to that of the proton, and comprising, along with the proton, all atomic nuclei. Free neutrons are unstable, decaying into protons, electrons, and neutrinos with an average lifetime of about ten minutes.

Nonabelian gauge theory: A different name for Yang-Mills theories that reflects the mathematical symmetry, called gauge invariance, that underlies them.

Non-Euclidean geometry: The specific application of Riemannian geometry to spaces that are not flat.

Parallax: The amount by which nearby objects, when viewed from different vantage points, will shift in comparison to distant background objects. The magnitude of this shift can be used to determine the distance of the nearby objects.

Parity: A parity transformation interchanges left and right. Certain interactions, like the electromagnetic interaction, do not distinguish between left and right. However, the weak interaction remarkably does distinguish between left and right, so that neutrons rotating around a certain axis will produce electrons that preferentially head off in one hemisphere, as opposed to the other hemisphere.

Photon: The elementary "quantum" of the electromagnetic field, a.k.a. light. Because of quantum mechanics, light has both wavelike and particle-like properties. In particular light of a given frequency is transmitted via many individual photons, so that for light of a low enough intensity, a detector will be able to detect the individual packets of energy carried by these particles, and never any smaller amounts.

Pions: Elementary particles produced in the collisions of energetic protons with matter. These particles, about ten times lighter than the proton, are made up of a quark and an antiquark, and are unstable with a lifetime of less than a millionth of a billionth of a second.

Planck scale: This is the length scale (or equivalently the energy scale) at which quantum mechanical effects relevant to gravity cannot be ignored. Because gravity is so weak at normal scales, it turns out that one must go to incredibly small scales before quantum effects become important. The Planck length scale is about 10^{-33}cm.

Precession: If a rotating or orbiting object returns to its initial position, and repeats precisely the same motion again, there is no precession. However, if upon returning to the same or position, the next orbit, or rotation, is shifted compared to the first, so that the motion does not exactly repeat after one such cycle, one says that the orbit or rotation is precessing.

Proton: An elementary particle with positive electric charge equal and opposite to that found on the electron. The proton, which weighs almost two thousand times as much as the electron, is located, along with neutral particles called neutrons, within the dense nucleus at the center of atoms. As far as we can measure, the proton is absolutely stable, but most grand unified theories predict the proton can decay with a lifetime too long to have yet been measured.

Quantum electrodynamics (QED): The theory that successfully combines quantum mechanics, relativity, and electromagnetism to correctly predict all phenomena that have been observed associated with the interactions of matter and electromagnetic radiation.

Quark confinement: The property that is associated with the fact that isolated quarks are not observed in nature. While this property of the strong interaction has not been mathematically proved yet, it appears to arise naturally as a corollary to the fact that the force between quarks gets weaker they get closer together, and stronger as you pull them apart. If this behavior continues indefinitely as you try and pull them apart, it would take an infinite amount of energy to produce a single quark, isolated from all its neighbors.

Riemannian geometry: A generalization of the flat two-dimensional geometric relations of Euclid, applied instead to spaces that can also be curved and that can also involve more than two dimensions.

Scattering: When two elementary particles collide together many different things can happen, from a simple grazing collision in which the particles are each deflected, to collisions in which the particles change their identities, and in which new elementary particles are created. All of these processes are called scattering processes.

Singularity: Generally describes the mathematical characteristic of any quantity that can grow infinitely large. When referred to points in space, a singularity refers to a region of space where the density of matter and energy grows infinitely large, and where the classical laws of general relativity appear to break down.

Spacetime supersymmetry: A mathematical symmetry that incorporates supersymmetry with the other known symmetries of space and time, including the fact that the laws of physics are unchanged from place to place, and from time to time.

Spacetime: The four-dimensional universe made up of three dimensions of space and one dimension of time, unified together by Einstein in his special theory of relativity, and first described by Hermann Minkowski.

Spectra: The set of colors of electromagnetic radiation emitted by different gases when you heat them up. Each element has a unique set of such colors that identifies it. The laws of quantum mechanics allow us to calculate the spectrum of light emitted by atoms, in agreement with observations.

Spontaneous symmetry breaking: This occurs when some symmetry of nature, such as left-right symmetry, is violated by the particular circumstances in which we find ourselves, but not by the underlying laws of physics that govern that situation. So, for example, while electromagnetism does not distinguish left from right, and electromagnetic interactions are those that are chiefly responsible for the makeup of material objects, I can nevertheless find myself standing next to a mountain on one side of me and an ocean on another side of me. In this case, I can clearly distinguish my left side from my right side. Such an accident of my particular circumstances represents an example of spontaneous symmetry breaking. Here is another one: Say you are having dinner at a large round table. After everyone sits down, every place-setting looks identical, and glasses are located both to the left and right of each person. Nothing distinguishes which glass is associated with which person until the first person picks up a glass. After that, the original symmetry is broken, and every glass is associated with a unique person.

Supergravity: Another name for local supersymmetry

Supersymmetry: A mathematical symmetry that relates elementary particles having different spin angular momentum. Specifically, supersymmetry implies that for all particles having integer spin (bosons) there should exist particles of equal mass having half-integer spin (fermions).

Tachyon: A hypothetical elementary particle which travels faster than the speed of light, and which can never be slowed down to below the speed of light. Such a particle could appear to an outside observer to be traveling backward in time. Generally, if a theory appears to predict tachyonic states, it is a sign that there is something unstable in the theory. Often such a prediction is associated with a violation of unitarity in the theory.

Tensor algebra: Mathematical relations that involve objects with multiple separate components, each of which can have a different dependence on both space and time.

Tesseract: A four-dimensional version of a three-dimensional cube. The "faces" of this hypercube comprise eight different three-dimensional cubes.

Torus: A donut-shaped object, with a hole in the center. One can produce such an object by taking a flat piece of paper and pasting together two opposite edges, and then the other two opposite edges. Alternatively, one can simply lay the paper flat and merely "identify" the two edges, so that for example, whenever an object heads off the right edge of the paper it would appear coming in from the left edge. In the language of topology, a torus therefore is topologically distinct from a flat piece of paper, in that it has a hole, but geometrically it can still be considered flat.

Uncertainty principle: One of the fundamental principles of quantum mechanics that implies there are certain combinations of quantities associated with any object that can never be measured exactly. For example, both the position and the momentum (see *momentum*) of an object cannot be known together with absolute accuracy. As one measures the position of an object more and more accurately, the uncertainty in knowledge about its momentum will decrease. Since this minimum combined uncertainty in position and momentum is, however, very small, the effect of the uncertainty principle is not usually directly observed on scales much larger than the size of atoms.

Unitarity: A fundamental mathematical property of nature that essentially says that probabilities do not change over time. Simply put, it implies that when one considers all of the different possibilities that may arise when one particles interacts with another, and sums up the different probabilities, they will add up to unity.

Vacuum energy: The energy associated with empty space, containing no matter or radiation. While common sense says that this energy should be zero, the laws of quantum mechanics and relativity together imply that empty space is full of a swarm of "virtual particles" that pop in and out of existence on a timescale so short we cannot observe them directly. When we try and calculate what the contribution of these particles might be to the energy of empty space, we come up with a very large number–indeed, far larger than anything we measure today. We currently do not understand why this prediction is incorrect. At the same time, any such energy, if it exists, is gravitationally repulsive, and could cause the observed expansion of the universe to accelerate. This is what we observe today in the universe on large scales.

Virtual particles: The laws of quantum mechanics and special relativity together imply that elementary particles and their antiparticles can spontaneously appear together out of empty space, exist for a short time, and then annihilate again, leaving nothing but empty space. As long as they do so for periods so short that we cannot measure them directly, their existence is ensured by the uncertainty principle. While virtual particles cannot be directly observed, their indirect effects can be observed, and predictions agree well with observations.

Vortex rings: A ring, like a smoke ring, that is stable and can move about, maintaining its form even as it moves through some background medium like air.

Warped space: This has, alas, nothing to do with *Star Trek.* Rather, it is a term that has been used to describe certain extra-dimensional theories with possibly large extra dimensions. In these theories the geometry (and hence the strength of gravity) in the three spatial dimensions we experience is not separated from the existence of the higher dimension(s), but is rather a function of where you are located in the higher dimension(s). In this case, it is possible not only for all familiar particles and nongravitational forces to be confined on our three dimensional space, but also for gravity to be effectively restricted to lie in our space, leaving the higher dimensions thus far undetected, but in fact allowing the possibility of their detection in new high energy accelerators such as the large hadron collider, and also allowing a possible new approach to the hierarchy problem.

Wavelength: For any periodic wave, with peaks and crests, the distance between successive peaks is called the wavelength of the wave.

Weak scale: This is the energy (or length) scale at which the weak interaction, responsible for the nuclear reactions inside the sun, for example, becomes of roughly comparable order in strength as the electromagnetic interaction, and which the mathematical symmetry between these two forces of nature, which is spontaneously broken at large scales, becomes manifest.

Yang-Mills theory: This represents a wide class of physical theories that are generalizations of electromagnetism, in which the particles that play the role of photons in electromagnetism, which are neutral, are instead charged, and also can have a mass, and therefore have more complicated interactions with one another and with other particles than photons do. Both the weak force and the strong force are described by Yang-Mills theories.

INDEX